Atmosphere in Urban Design

T0341129

This book offers an ethnographic exploration of the role that atmosphere plays in work processes undertaken within an urban design studio. It provides understandings of how architectural practices are fuelled with atmosphere in various configurations throughout different design phases of selected projects for construction. From the outside architectural practices commonly appear well-ordered and carefully considered, established by proof and rationally justified. This book though poaches on architects' preserves in order to draw attention to features of unpredictability and uncertainty within the design phases. By opening up into the 'machinery room' of urban designers, the goal is not to spoil the plaster saint cover of a 'starchitect' business, but to remind about the crucial value that pockets of doubt issuing questions rather than answers, open-mindedness instead of single-mindedness, play to the processes of design production and creativity. The book identifies these pockets as atmospheres enveloping the architectural practice.

Anette Stenslund is an assistant professor at Southern University of Denmark. She was working as a post.doc. at Roskilde University, Denmark, while this book was written.

Ambiances, Atmospheres and Sensory Experiences of Spaces

Series Editors:

Rainer Kazig, *CNRS Research Laboratory Ambiances – Architectures – Urbanités, Grenoble, France*
Damien Masson, *Université de Cergy-Pontoise, France*
Paul Simpson, *Plymouth University, UK*

Research on ambiances and atmospheres has grown significantly in recent years in a range of disciplines, including Francophone architecture and urban studies, German research related to philosophy and aesthetics, and a growing range of Anglophone research on affective atmospheres within human geography and sociology.

This series offers a forum for research that engages with questions around ambiances and atmospheres in exploring their significances in understanding social life. Each book in the series advances some combination of theoretical understandings, practical knowledges and methodological approaches. More specifically, a range of key questions which contributions to the series seek to address includes:

*   In what ways do ambiances and atmospheres play a part in the unfolding of social life in a variety of settings?
*   What kinds of ethical, aesthetic and political possibilities might be opened up and cultivated through a focus on atmospheres/ambiances?
*   How do actors such as planners, architects, managers, commercial interests and public authorities actively engage with ambiances and atmospheres or seek to shape them? How might these ambiances and atmospheres be reshaped towards critical ends?
*   What original forms of representations can be found today to (re) present the sensory, the atmospheric, the experiential? What sort of writing, modes of expression or vocabulary is required? what research methodologies and practices might we employ in engaging with ambiances and atmospheres?

The Aesthetics of Atmospheres
Gernot Böhme. Edited by Jean-Paul Thibaud

For more information about this series, please visit: www.routledge.com/Ambiances-Atmospheres-and-Sensory-Experiences-of-Spaces/book-series/AMB

Atmosphere in Urban Design
A Workplace Ethnography
of an Architecture Practice

Anette Stenslund

Routledge
Taylor & Francis Group

LONDON AND NEW YORK

First published 2023
by Routledge
4 Park Square, Milton Park, Abingdon, Oxon OX14 4RN

and by Routledge
605 Third Avenue, New York, NY 10158

Routledge is an imprint of the Taylor & Francis Group, an informa business

British Library Cataloguing-in-Publication Data
A catalogue record for this book is available from the British Library

Library of Congress Cataloging-in-Publication Data
Names: Stenslund, Anette, author.
Title: Atmosphere in urban design : a workplace ethnography of an architecture practice / Anette Stenslund.
Description: Abingdon, Oxon : Routledge, 2023. | Includes bibliographical references and index. |
Identifiers: LCCN 2022021050 (print) | LCCN 2022021051 (ebook) | ISBN 9781032247090 (hardback) | ISBN 9781032247106 (paperback) | ISBN 9781003279846 (ebook)
Subjects: LCSH: Architecture and anthropology. | Architectural practice.
Classification: LCC NA2543.A58 S74 2023 (print) | LCC NA2543.A58 (ebook) | DDC 720.1/08--dc23/eng/20220623
LC record available at https://lccn.loc.gov/2022021050
LC ebook record available at https://lccn.loc.gov/2022021051

ISBN: 978-1-032-24709-0 (hbk)
ISBN: 978-1-032-24710-6 (pbk)
ISBN: 978-1-003-27984-6 (ebk)

DOI: 10.4324/9781003279846

Typeset in Times New Roman
by KnowledgeWorks Global Ltd.

For Esther Eo

Contents

Images

Acknowledgements

My gratitude goes to SLA and in particular to Stig L. Andersson and Alexandra Vindfeld Hansen for welcoming my presence and supporting my investigations throughout the ethnographical fieldwork period. A tremendous thank you to all employees who shared with me their working days, initiating me into their duties and willingly sharing their experiences, thoughts and concerns regarding whatever I would ask about. The studio as a whole provided the ethnographical foundation of this book.

My position at the Department of People and Technology at Roskilde University Centre and my affiliation with the *Living with Nordic Lighting* team of researchers led by Mikkel Bille and funded by Velux Foundation [#16998] has provided me with stimulation and a collegial environment and I am grateful for the institutional support. My dear colleagues within the team deserve my warmest acknowledgements. Thank you to Mikkel Bille for leading the team, and for patiently supporting, cherishing and at times pushing me through the process of writing this book. Mikkel has been invaluable for keeping me on track in many of the book's chapters. I have learned a lot about overview, productivity and persistence in Mikkel's company, for which I cannot thank him enough. I am grateful to Mikkel, David Pinder, Siri Schwabe, Jeremy Hektor Payne-Frank, Ida Lerche Klaaborg and Olivia Norma Jørgensen for being the good colleagues that they are. I thank them for early feedback on several chapters for this book and for regular food for thought, such as during our 'Thinking Thursday' evenings before Covid-19. Also, I wish to thank members of MOSPUS – the interdisciplinary research group in Mobility, Space and Urban Studies at Roskilde University Centre – for stimulating conversations.

Friends and family have seen a little less of me during the most intensive periods of the writing process – thank you for your patience. Heartfelt thanks to Nanna Greiersen for silhouetting images. Thanks also to my friend Ken Arnold who has supported me with diversion but also for reading and commenting on early drafts of selected chapters.

I dedicate this book to Esther Eo. She was seven months old as her mother started the investigations that gave rise to it, and as it goes into print she will be four years old. No words suffice for the rapture of watching her grow up.

1 INTRODUCING ATMOSPHERES IN URBAN DESIGN

Atmospheric design is itself the product of a particular atmosphere
—Mark Wigley (1998: 26)

Atmospheres may be designed. Spaces are orchestrated according to scale, objects are shaped and placed, lights and sounds arranged and surface textures meticulously selected in order to provide urban areas under redevelopment with the desired vibe. Yet no designer is in control of the appearance of an atmosphere as it emerges. Whether designers market atmospheres or choose instead not to make them part of their services, atmospheres will be present as by-products regardless. Thus, disobedient, they intrude on the very design processes, capture the designers themselves and ultimately shape the life circumstances of ordinary people inhabiting designed spaces. Whether we like it or not, atmospheres continuously show their presence, and since we cannot get rid of them, this book suggests that the industry comes to terms with them and that the designers learn to make profit from their potential, which calls for more knowledge about how atmospheres emerge in urban industries.

This book furthers the understanding of atmospheres in urban design. It shows how atmospheres are inevitably *in* urban design, and it discusses how attempted urban atmospheres are evoked, manufactured and represented *by* design in various ways. In brief, the book is concerned with demonstrating, discussing and analysing how atmospheres flourish *within* the work processes in which urban designers engage. Such atmospheres are not yet 'out there' in urban space, they are contained *in* situations as they play out in the design offices – in front of a computer screen, in the interchange with software, co-workers, drawings and materials. The book thus shows ethnographically how atmospheres envelop and enliven urban redevelopment and demonstrates how a good many such envisioned atmospheres arise from situations of uncertainty that are key to the designers' creative output.

The intention goes beyond simply acknowledging the fact of designers' commitment to uncertain futures (Akama et al., 2018: 10; Light, 2015; Smith et al., 2016) that move beyond the knowable in order to change present

DOI: 10.4324/9781003279846-1

situations into imagined and preferred ones (Simon, 1968). Aligning such insights with understandings of atmosphere allows us to consider what it might entail when designers engage their feelings and bodies in the uncertainties inherent in the processes of designing – what costs it has and what benefits it brings. As Akama et al. write, uncertainty 'can bring forms of discomfort but it can also inspire and [...] open up new possibilities' (2018: 126). This book explores the possibilities through the study of atmosphere while still acknowledging vocational requirements to curb it (Samimian-Darash and Rabinow, 2015). In brief, the argument I put forth is that urban design feeds on atmospheres from within and throughout the developing processes that are less evident to an outside world. To perform well and devise solutions of a desired quality, designers will inevitably engage in atmospheres, which means that they will resonate with their production in order to *feel* it. Designers cannot take control of the designed atmospheres, but as I will show, they are *with* their design in reciprocity – in atmosphere.

'Everything can be measured. Everything', the Head of the Danish Association of Architectural Firms (colloquially known as Danske Ark) reminded me in an interview. Their Chief Consultant, present at the same interview, added that what Danske Ark see as their mission is to turn all 'soft values' into 'hard facts'. On that note, on their homepage and in a publication Danske Ark highlight selected model examples of architectural teams that profile their activities with claims to be architecting happiness, health, comfort, safety, learning, social cohesion, etc. (Sejr, 2017). No transparency is offered by any of the projects mentioned about how calculations support the assumptions. Simple counts of peoples' movement patterns may be seen as a token for the physical environment's direct effect on complex social conditions. For example, attractive newly erected library buildings and education and sport centres whose increased visitor numbers seem to reflect approval are said to enhance learning, health, community and ultimately provide economic value.

The examples reflect a general picture of a ruling attitude within the architectural industry privileging matters of fact and diluting or even forcing the more complex, experiential qualities of urban design into 'clear', measurable data. There is a tendency to prove a certain effect of designed environments, but I wonder whether the market is not sophisticated enough to know that social conditions cannot shut off uncertainties and thus be calculated with the same success as can economic, material and physical environmental aspects. For example, anyone who has been a teacher or student will know what people might occasionally learn in given situations is quite a complex matter not adequately documented by recording behaviour alone. Pupils can show up without being present, that is, being absent in mind while physically present, and whether users of sport centres are healthy and community-oriented is also not easy to answer unequivocally – such claims would require thorough investigation. The examples illustrate how urban designers' claims can weaken, phrases can become meaningless, and in the worst-case scenario, companies' reasoning becomes mistrusted, and their credibility is questioned when future complex matters are addressed with opinionated certainty.

No single architectural project highlighted by the Danish Association of Architectural Firms discusses how people feel, 'find themselves in' or experience their surroundings – not even in a publication about architectural value (Sejr, 2017). This question about the feel of urban design is of concern to the atmosphere approach developed within this book, and the experiential perspective is the pivotal point for any of its discussions. To be transparent about the situated nature of my authorship, as a sociologist I am trained in producing and analysing quantitative and qualitative data on an equal footing. To clear up any potential misunderstandings: this book is not against either measurements or certainty. Yet I see the rational wave of expectations to scientifically state facts captured in design initiatives as a problem precisely because it threatens the credibility of the architectural industry. We might sensibly count people's behaviour, detect their movement patterns, monitor their palpitations and measure their production of hand sweat. There is ultimately no limit to what we can measure, and measures surely give *some* indication of the effects of architecture and landscape design. However, most people will also know that how happy, sad, healthy, wise or safe people *are* in given situations depends on how they *feel*, and complexities and nuances can rarely be measured.

I am not the first to draw attention to the importance of atmosphere in this respect and to the problems of quantitative ways of evidencing the impact of spatial design. In the anthology *Designing Atmospheres,* landscape architect Weidinger writes: 'Unlike functional considerations, qualitative phenomena are not measurable. Their description and evaluation require a different approach' (2018: 9). In my opinion, integrity is required about the scope of evidence, whether quantitative or qualitative. Since not all matters in life can be explained and foreseen in a clear-cut factual manner, one way to gain trust in the industry, I suggest, is via the recognition of atmospheres characterised by uncertainty and emergence as the foundation of design practices. Before I elaborate and situate the literature from this perspective, I will explain my understanding and use of the concept of atmosphere.

Atmosphere

The concept of atmosphere is far-reaching. It may refer to meteorological conditions, such as the vapour enveloping the earth (Andersson, 2018: 115; McCormack, 2014), the air we breathe and exhale (Anderson, 2009: 78), smell (Stenslund, 2015: 347; Stenslund, 2017: 157; Tellenbach, 1968: 46) or the 'spatial feeling of being attuned in and by a material world' (Bille et al., 2015: 35). Atmosphere may also be described as a 'spatial phenomenon' (Sumartojo and Pink, 2018: 19) that 'folds together affect, emotion and sensation in space' (Edensor, 2015: 83), thus imparting certain characteristics or intensities to specific geographical locations. Think of the solemn atmosphere of a church, the peculiar atmosphere in a museum, the susceptible atmosphere of a hospital, the relaxing vibe of a public park, the cosy atmosphere of one café over another, etc. While atmospheres are felt, sensed and thus 'often understood in spatial terms, they are not limited by particular spaces – that is, they should not be thought of as bound or contained by space, or beginning and ending in clearly identifiable ways' (Sumartojo and Pink, 2018: 56). I concur with German phenomenologist Hermann Schmitz that atmospheres are spatial but placeless feelings that mark the body. In his words: 'Feelings are spatial, but placeless, poured-out atmospheres' (2009: 23, my translation) – 'and corporeal capturing forces' he adds elsewhere (2014: 30, my translation). Like clouds (Rauh, 2018: 43), atmospheres are in a constant flux; they change and reshape continuously as they depend on the situations (Riedel, 2019: 173) played out between people and their material and sensory surroundings.

Atmosphere in this book is thought of as shared feelings and moods linked to situations and physical surroundings in which bodies – human and inhuman – take part. Atmospheres, therefore, are feelings that do not belong to individual beings; they are not subjective; they cannot be ascribed to 'inner' mental states; and they are never owned by private 'selves'. The quotation marks indicate the social construction of such inventions and mirror the aim of subverting binary distinctions between subject and object, inner and outer matters. Atmospheres assume an ontological starting point that recognises feelings as something that flourishes 'in the air' to be jointly embodied and culturally inflected. For example, my private mental state does not alone determine the experienced character of my spatial surroundings; it can certainly colour my impression and infect other peoples' impression but there is no absolute projection taking place. The culturally situated space, with its present materials, materialities and potential presence of others, will inevitably influence my state of being. The situated space is always already loaded with atmosphere, and as another German phenomenologist, Otto Friedrich Bollnow, says: 'it can be cheerful, light, gloomy, sober, festive, and this character of mood then transfers itself to the person staying in it' (2011: 217). Man and space 'belong together' in atmospheres: 'One speaks of a mood of the human temperament as well as of the mood of a landscape or a closed interior space, and both are, strictly speaking, only two aspects of the same phenomenon' (Bollnow, 2011: 217). Hence, atmospheres are shared: they do not belong to anybody, and they cannot be anchored; we know them as shared experiences or common denominators. Schmitz says that 'atmospheres are super-personal, joint feelings […], the "mood bell" that unites all present' (Schmitz, 1996: 53, my translation). Mood, in German understood as *Stimmung*, I refer to as shared feelings, and spatially and culturally situated shared feelings are atmospheres.

While there may be a point in separating the two concepts (Pile, 2010; Seyfert, 2012), in the book I refer to feelings and emotions interchangeably.

Phenomenologically I approach atmospheres in urban design from a life-world dimension that, in line with what philosopher Tonino Griffero calls 'pathic' aesthetics, 'defends the validity of sensible-emotional experience and resists the attacks from scientific abstraction, thus highlighting the qualitative involvement in the world' (Griffero, 2019: 414). Aesthetics, here, lead well beyond artistic and cultural understandings of beauty, with epistemological reference to the Greek *aistheses* that involves an emotional kind of embodied sensing of what with Griffero can be termed the 'sensible-emotional' with a hyphen that stresses how sensory perception and feelings go together and are not separable.

Although 'affect' is a fairly common term in English, I deliberately avoid using it. The reason is that in an academic context, affect theory has opened ways for non-representational branches of theory to engage with 'affective atmospheres' (Adey, 2014; Adey et al., 2013; Anderson, 2009), and such couplings of affect and atmosphere come with a risk of understanding atmospheres by proxy (Bille et al., 2015: 35) as well as underestimating how atmospheres are as much 'done', 'performed' and 'practised' by human living bodies as they are evoked by material qualities (Bille and Simonsen, 2021). I have previously directed a similar criticism at material studies that tend to assume a physical presence of bodies in order for atmosphere to be generated, experienced or shared (Stenslund, 2014: 73). An example is given by Teresa Brennan's vivid book *The Transmission of Affect* wherein she argues that affects contagiously spread from person to person via neurological entrainment and hormonal interaction (2004: 9), thereby turning atmosphere into as much a biological and physical effect as a social phenomenon. There is nothing wrong with biological, physiological or material approaches, in fact they are most relevant as long as atmosphere is not reduced to the presence of physical matter. Affective atmosphere, however, risks losing touch with everything that is culturally driven, felt and experienced and everything that is absent (*not* materially present) yet can hold a crucial atmospheric presence by, through mood, being able to touch and involve our state of being.

Another critical move I concur with is affect theories' analytical anticipation of causal explanations of atmosphere: 'Seemingly ephemeral, seemingly vague and diffuse, atmospheres nevertheless have effects and are effects' (Anderson and Ash, 2015: 35). But urban design described in terms of cause and effect, I counter, requires an analytical approach that separates designed environments from their creators and users in order to detect directional power relations typically travelling from the designer's mind into the material design and further on to the user's wellbeing. This approach is far from that of phenomenological ambitions to overcome dual preconception, and in this sense, there are clear differences between affect and atmosphere. Friedlin Riedel understands the difference between affect and atmosphere as follows: 'While "affect" refers to the ways in which (emerging) bodies relate to each other [...], "atmosphere" describes the ways in which a multiplicity of bodies is part of, and entrenched in, a situation that encompasses it' (2019: 85). With such a view, the difference seems to be given between a relational and a holistic perspective. To me, affective atmospheres evoke an image of units behaving like billiard balls colliding with various (measurable!) forces, whereas phenomenological atmospheres (that are dealt with in this book) are not grasped through difference, separation and distinction but rather from a perspective that allows beings to unite, merge or interdependently work together.

Given the different conceptual understandings, I usually like to think of atmosphere unfolding in the space *in between* (a) perceiving human bodily beings, (b) the sociocultural situation played out and (c) the physical surroundings, no matter how much they are touched by human hand, and no matter how 'natural' or artificial they are considered to be. This space in between is an experienced subset that cannot be escaped: we are always situated and feel atmosphere whether it is boring, trivial or awfully exciting.

Sometimes urban designers speak about 'the public realm' or 'the life between the buildings' (Gehl, 1987), but atmosphere offers a different lens through which to also attend to the invisible and immaterial experienced value (not effect) of urban design. For atmospheres concern feelings that are less easy to observe, pinpoint and register objectively, yet they are not of private subjective concern. Some feelings are out there floating in the open, dependent on our engagement with the surrounding world, and sometimes such feelings can overwhelm. In such situations we give in to atmospheres, and we might come to experience how atmospheres change or deeply transform our previous mood (Griffero, 2021: 36). This is what gives atmospheres their 'pathic' character: they can appear possessive and, accordingly, receptive people may 'act only in the light of this rapture' being *subject to* atmospheres (Griffero, 2019: 415 with reference to Ludwig Klages, 1974).

Atmosphere in processes of making urban design

A significant literature on atmosphere, architecture and urban design is characteristically interested in understanding how architecture is perceived from a user's perspective (Bille and Sørensen, 2007; Böhme, 2014; Borch, 2014; Hasse, 2014; Kazig, 2008; Pallasmaa, 2014; Thibaud, 2015. Informed by this literature we can understand the experienced, felt and lived quality of urban design, and how urban spaces always touch atmospherically in ways that will never be exhaustively addressed by rational, scientific statements (Hasse, 2014: 203). This knowledge informs several exciting ethnographic enquiries on atmosphere and architecture today that primarily aim to explore atmospheres as they play out in urban and rural spaces, at events, in homes, hospitals, museums, pubs or other situations and settings (Bille, 2020; Edensor, 2012, 2016; Schroer and Schmitt, 2018; Schwabe, 2021; Stenslund, 2015; Sumartojo et al., 2019).

When it comes to understanding *the processes of making* in terms of atmosphere, however, the literature is rather thin. Atmospheres can be produced, Böhme suggests (1995: 16), but what exactly this might imply, and also what actually goes on in and around the architectural offices that claim to either create or escape atmospheres, is no more than hinted at. Donald Schön and Chris Argyris were among the first to undertake observational studies within design studios. In doing so, they highlighted the difference between espoused and enacted practices of practitioners including architects, engineers and industrial designers (Argyris and Schön, 1974). In addition, Schön approached the epistemic practices of designers through these encounters on-site in the offices (Schön, 1979, 1985, 1987). With her ethnographic exploration of socially constructed and negotiated architectural practices, however, it was Dana Cuff (1992) who pioneered long-term workplace ethnography within this field. Through six months of observational studies divided between three projects in San Francisco, she sounded out the culture of social architectural practice as it was 30 years ago.

Building on the growing interest in understanding how experts learn and know about their subject fields of knowledge, Albena Yaneva carried out long-term ethnographic fieldwork consisting of three years of meticulous observations tracing the development of a single project – the development of the Whitney Museum of American Art in New York by the Office for Metropolitan Architecture (OMA) of Rem Koolhaas in Rotterdam. Following the Actor Network Theory (ANT) as a method in Science and Technology Studies, Yaneva literally observes how designers move, gesticulate and react to buildings, mock-ups and measurements that 'lend themselves to interpretations' (Yaneva, 2009: 7).

Despite their widely differing approaches to examining urban design processes, I read both Cuff and Yaneva as confirmations of my view that the notion of emergence and uncertainty is crucial to the making of design. Cuff shows clearly that design processes are social processes, and due to their high degree of social complexity caused by many participants (clients, collaborators, citizens, suppliers, etc.), they become rather opaque and difficult to predict and foresee even by the participating designers themselves – an insight that informs current research in collaborative design (co-design). Following the perspective given by Yaneva, in turn, the design processes (including the 'practices', as she prefers to say) appear to be beyond the designers' control for the reason that they are conceived as 'experimental' by nature. For example, construction work can resist attempts at control and act 'disobediently', going against the designer's original plan (Yaneva, 2008). In the aforementioned pioneering ethnographies undertaken within architectural design offices, there is no mention of atmosphere or felt sensations, but more recent studies of urban design resting on the ANT have made significant advances suggesting for instance the designer's visual production of 'digital atmospheres' (Degen et al., 2017: 9) through digital visualisations that 'emphasise the atmosphere' (Rose et al., 2016) or 'the atmospheric evocation of what it would "feel" like to be in' given places under development depicted through image-making (Melhuish et al., 2016: 229). Likewise, in Houdart and Chihiro's ethnographic study of the Japanese architect Kuma Kengo's work practices (2009), we are given slight indications of designers' work with atmosphere. Admittedly, the focus of their monograph is not on atmosphere, and it does not rest on the ANT either, but *en passant* it mentions how Kuma Kengo engages with the construction sites by way of feeling their atmosphere (2009: 49), and also how his team of employees seeks to create 'Japanese atmosphere' in a project (2009: 107), which all adds to the literature on the production of atmosphere.

Atmosphere in Urban Design is in conversation with all the authors mentioned, but it differs significantly by having a pronounced interest in how atmospheres bring together urban designers and their design objects in various activities. Hence, it adds to the existing body of literature with a felt and embodied perspective. A recent ethnographical work carried out by anthropologist Thomas Yarrow (2019) in the office of Millar Howard Workshop (MHW) in Stroud, in the county of Gloucestershire in England, deserves particular mention because it attends to the emotional commitment to the design process itself, which resonates with the observations I myself have made. In his vivid narrative portrait of architects at work, Yarrow not only mentions how architects develop a 'feel for place' by 'just taking things in' through their site visits (2019: 76) but also that the design process itself involves feelings. Yarrow is not concerned with atmosphere as such, and in general he is determined to 'refuse' and 'scale back' any kind of theoretical exegeses in favour of a descriptive presentation of 'architects' own explanations' (2019: 6). Nevertheless, he talks about conversations with his informants (architects) that seem to confirm the present book's argumentation about atmospheres that feed the very processes of making.

For example, Yarrow touches on the topic of architects' relationship to uncertainty. As he asks them where their designs come from, the architects 'confess their own uncertainty about the process that is both familiar and mysterious' to them (2019: 110). 'Total magic', one informant asserts, while another starts to describe how 'the design acquires an energy that [...] the architect "feeds off" [...] Design starts to become an entity before it becomes a building. The architect can "guide it," "steer it," "feed it with energy," but does not in any straightforward sense control it', Yarrow learns (2019: 111). Accordingly, he traces the architects' role as one that responds to and acts 'as a kind of receptor' to what 'the design itself is telling':

> Design, as they [architects] see it, is an action that they produce and a way of responding to the actions of the objects that result. The process involves oscillation: between a self that is 'in control' and one that is 'led'. At times this may blur their sense of self, of who they are in relation to what they have made [...].
>
> (Yarrow, 2019: 112)

Although Yarrow argues that individuality remains central to some phases of urban design, I partially oppose that argument proposing that the 'respondence' and 'oscillation' actually *is* atmosphere. Atmospheres dissolve the idea of a designer subject and a design object as separate entities, and the separation needs no theoretical recovery as Yarrow seems to suggest (2019: 112). Phenomenologically, we are always already 'out in' the world: 'in this living-towards, there is something of me: my "I" goes out beyond itself and resonates with [it]', Heidegger writes (2000: 62). New insights might therefore follow if we embark on a phenomenological approach acknowledging how the designer and the designed might be interconnected in various ways – in atmosphere. Phenomenological sensibility, according to Hermann Schmitz, is all about the interconnectedness of 'I and world' (1978: 15). From this point of view, designers are no longer situated in a relation at arm's length from what they make but are conceived *as* the relation as such – that is, one that exceeds individual and objective bodies. Individuals may at times claim their independent creativity (cf. Yarrow, 2019) just as they may at times argue for their interconnectedness (cf. Cuff, 1992). Whatever they do, they find themselves in atmosphere – they are, with Böhme, 'exposed to the world and oscillate with what is perceived' (Böhme, 2001: 83, my translation). 'Atmospheres are something between subject and object. They are not something relational, but the relation itself' (Böhme, 2001: 54, my translation). The book will support this argument empirically.

There are a few articles written by designers and architects themselves about the rules and tools for designing atmosphere (Weidinger, 2018). The responses given to the question of how to work with atmosphere in design are, however, a world away from the proposals offered in this book. Espoused practices often turn into ideology, or they serve as sales pitches in disguise transformed into essayistic, poetic and philosophical writing on atmosphere ontology – Zumthor's essay on atmosphere (2006) is a model example. Whenever designers talk about atmosphere, they tend to do so as something positive and desirable, like epiphanies that reveal themselves to the designers' surprise as well. This observation affirms what Griffero (2014) has already assumed theoretically, that it is often seen as a compliment if we talk about something having atmosphere. In the next chapter I elaborate on the methodological implications of architects' more exclusive understanding of atmosphere, but here it is enough to point out that atmosphere clearly becomes a marketable service for the industry, and that the chapters in this book will reflect this tendency to associate good quality design and successful services with atmosphere. As an ethnographer I am primarily interested in the enacted practices and processes (what designers in fact do) when working with atmosphere and less interested in the espoused practices (what they would like to do) or how they strategically promote themselves and brand their strategies, flagging the phenomena of atmosphere in design. The methodological implications of the ethnographer's efforts to dig beneath the official stories of design businesses is discussed in the next chapter, dealing with methods and methodology. Below is a brief introduction to the company I worked with. A fuller 'portrait' of their espoused and enacted practices is provided in Chapter 3.

The SLA studio

The book draws on nine consecutive months of ethnographic fieldwork in SLA – a renowned urban design studio consisting of two offices in Denmark (Aarhus and Copenhagen) and one in Norway (Oslo). SLA takes its name from the initials of Stig L. Andersson, who founded the company in 1994. Today, SLA employs 130 people covering more than 15 nationalities, and their interdisciplinary team comprises landscape architects, biologists, anthropologists, city planners, lighting designers, architects, planting experts, urban designers, ecologists, safety experts and forest engineers. While a large market share is within Scandinavia, their global impact is constantly growing with completed projects in Europe, Asia, the Middle East, Africa and North America. Interdisciplinarity is paramount at SLA and when I ask the employees what they do, most people say they do nature-based urban design. Andersson agrees and nuances that, to him, nature-based design is art: 'We make art', he says. 'It may be hard for some to understand, but that's what we're doing. Our art is the nature-based design'. In Chapter 3, I describe the ideological foundation for this design and the principles I have been able to observe for its execution – the *espoused theory* claimed, and the *theory-in-use* solving problems for the clients (Argyris and Schön, 1974).

With one exception, all informants mentioned, paraphrased or quoted in the book are anonymised. The exception is Stig L. Andersson, the founder of SLA, who today acts as design director and partner. Andersson is difficult to make anonymous because he remains a leading figure at SLA and many employees refer to him – by first name as will appear when I quote them – to explain themselves when talking to me. Anonymity is also maintained in the emails reproduced in Chapter 5. Company and project names are real, but email addresses (except for my own) and names are fictitious throughout.

In accordance with SLA, who tell me how much they value and cherish their informal studio-character, to avoid being repetitive I sometimes refer to the company as 'the studio'. Out of respect for the many disciplines that today contribute to SLA's nature-based design, SLA has expressed a wish not to be referred to as an architectural firm or landscape architectural company. However, it is difficult in a book like this to completely avoid reference to architects and architecture or landscape architecture, so when that happens there are no intentions of highlighting one profession's skills over others, but only a need to adjust to the common terminology used by associations of firms engaged in urban design at large. Furthermore, with regard to my written language, I often employ and import the vocabulary of informants themselves into my own conceptualisation, which is a data processing strategy inspired by 'in vivo coding' (Glaser and Strauss, 1967). Hence, I very often make use of words that are not my own but come from the everyday formulations of my informants. These 'in vivo words' appear in quotation marks.

The outline of the book

Chapters 1–3 set the scene for the following series of chapters that analytically conceptualise and discuss the empirical information derived from the ethnographic fieldwork in SLA. In Chapter 2, 'Investigating Atmospheres – Methods and Methodology', an account is offered of the methods used during the fieldwork and the continued cooperation which existed up to the publication of this book. The chapter reflects on the procedures, and their epistemic framing, of the research on which the book is based.

'Who are they?' a friend asked me, as I told him about this book that I was writing about the atmospheres enveloping a company's design processes. Chapter 3, 'Naturalising Atmospheres the Nordic *Wabi Sabi* Way', seeks to provide the reader with an answer to this simple but highly relevant question. It explains SLA's ideological foundation in the theory of complementarity and their espoused understanding of nature that is enacted in nature-based design. Based on observations of designers' practices and in conversation with them, I suggest the resonance with *wabi sabi* in Japanese life philosophy enabling me to identify a particular kind of culturally informed aesthetic taste at SLA – that is, a cultural value that determines the appreciation of a specific kind of designed nature.

Chapter 4, 'Collaging Atmospheres – Epistemic Spatial Physiognomy', starts by highlighting the epistemic value of the collage, which instead of closing down uncertainties, cultivates them. Beyond helping the designers to remain investigative, questioning instead of answering, the chapter suggests how collaging atmospheres relies on skilled approaches to landscape physiognomy, based on the designer's ability to carefully select, take apart and combine anew within the collage the key materials of a space that can make even an outsider feel its vibe. Collages, most commonly prepared in the early phases of the design, are described as artistic sanctuaries held at a temporary distance from the many demands and wishes from an outside world. Collaging thus occasionally becomes small pockets of artistic oeuvre that are no longer only produced to serve the customer but are largely prepared for the designers themselves – to distinguish one's business approach. Chapters 4 and 5 are the two chapters of the book which have been published elsewhere. Chapter 4 on collaging was first published in a slightly different version in *Environment and Planning B: Urban Analytics and City Science* (Stenslund, 2021).

Chapter 5, 'Rendering Atmospheres through Lighting Aesthetics', was originally written together with Mikkel Bille and published with a similar but not identical title in the anthology *Architectural Anthropology* (Stenslund and Bille, 2022). For Chapter 5, I edited and expanded on the first version mainly by reinserting some of the empirical findings and my original ideas that had to be left out of the shortened anthology version. The aim, however, remains the same, namely, to demonstrate how the perceived atmosphere of a place is communicated and rendered among urban designers, external collaborators and graphic visualisers. Urban designers may have an idea about how a place should *feel*, but how is this feeling communicated in renderings to make others feel the same or at least understand the vision for the atmosphere? The chapter illustrates how something as visceral as atmospheres is communicated and negotiated through the initial sketch proposal phases of a construction project. By following the process of making renderings, the chapter looks at how the designers communicate phenomena such as material elements, light and shadow in order to visualise atmosphere. The work is characterised by skilled ways of communicating about the intuitive flair of cultural aesthetics, where commonly used yet vague words are understood in rather precise ways among designers.

With Chapters 6–8, the book moves from the initial design phases (the sketch proposals and preliminary design) into the phases where designers draw in detail solutions for construction and submission of tender. But even at stages where uncertainty could be expected to turn to certainty, the observations show how, in parallel with the specifics and details of solutions, there is a continued open approach to atmospheric pockets of uncertainty that favour the *wabi sabi* aesthetic marking of the design.

In Chapter 6 on 'Scaling Atmospheres – Intuitive Digital Zooming in and out on Materiality', conventional understandings of scale involved in the quantitative approach to space geometry are reshuffled by the introduction of atmosphere in digital scaling. The chapter suggests how scaling designers engage with empathy in imagined future redeveloped spaces that, via scaling, are sought to be integrated into visualisations that resemble the renderings. A quality stamp for digital scaling in the studio is that it is based on intuition – knowledge instantly assigned by feelings rather than rational facts. Scaling is therefore just as much guided by atmosphere used as a tool in the studio as it is by the imagined atmospheres in future urban space outside the studio. Consequently, upscaled and downscaled visualisations serve not only to captivate users, clients or stakeholders of future urban spaces but also to enchant the design process itself, captivating and making the designers feel connected. The chapter thus calls attention to how quantitative and qualitative elements of design are infiltrated in the in-house digital scaling of atmospheres that provides a common cause for a project team.

Through the analysis of four examples of SLA's design practice, Chapter 7, 'Astonishing Atmospheres – Mimetic design', considers the complexity of a design process that mediates mimesis and alterity in various ways in order to produce exciting, enchanting, appealing, alluring urban design – varieties of experienced amenity value that I group together as an ideal type of astonishing atmospheres. Through the ins and outs of the creation of atmosphere, what appears to sustain the appeal of the design is mimesis that maintains a dialectic between sameness and alterity. Hence, the chapter shows how pure mimesis, as in copying, easily has a corny outcome that loses the atmospheric allure. This is where the designer plays an important role in their approach to the work process, attending to uncertainty in various ways, such as when they themselves allow the atmosphere of the design to be sensed, whether natural or artificially produced, fake or calculated in origin.

In Chapter 8, 'Drawing Atmospheres – Resonance in Digital Graphics', thorough observations are presented of how atmosphere envelops the digital drawing of future space. It demonstrates how atmospheres are neither projected from a designer's predetermined mind onto the design nor sent onto a receptive designer by the designed things articulating their presence through qualities conceived of as what Böhme calls 'ecstasies' (2017: 19). By pursuing the drawing of a cladding for a sheet pile wall, the chapter goes into detail with the atmosphere that envelopes the designer's way of resonating with the many parts of the design – including the design narrative, geometric measurements, shapes of profiles, light and shadows. The drawing process is characterised by a form of correspondence, but the correspondence transcends a physical and material exchange. It is characterised by a felt and sensed relationship best understood as resonance loaded with moods as atmosphere. Digital graphics draw in the designer and the designed into a resonating relationship. The last chapter thus draws attention to the value gain of designers drawing *with* the design, while resisting overthinking and oversteering it.

Beyond bringing together the chapters' findings in Chapter 9, 'Conclusion', the book is inserted into a broader context that reflects on how the surrounding world might meet atmospheres in urban design.

References

Adey P (2014) Security atmospheres or the crystallisation of worlds. *Environment and Planning D: Society and Space* 32(5): 834–851.

Adey P, Brayer L, Masson D, Murphy P, Simson P and Tixier N (2013) 'Pour votre tranquillité': Ambiance, atmosphere, and surveillance. *Geoforum* 49: 299–309.

Anderson B (2009) Affective atmospheres. *Emotion, Space and Society* 2: 77–81.

Anderson B and Ash J (2015) Atmospheric methods. In: Vannini P (ed) *Non-Representational Methodologies: Re-envisioning Research*. Abingdon: Routledge, pp. 34–51.

Andersson SL (2018) Atmosphere: A thin film of enclosure. In: Weidinger J (ed) *Designing Atmospheres*. Berlin: Technishe Universität Berlin.

Akama Y, Pink S and Sumartojo S (2018) *Uncertainty and Possibility: New Approaches to Future Making in Design Anthropology*. London: Bloomsbury Publishing.

Argyris C and Schön DA (1974) *Theory in Practice: Increasing Professional Effectiveness*. San Francisco: Jossey-bass.

Bille M (2015) Hazy worlds: Atmospheric ontologies in Denmark. *Anthropological Theory* 15(3): 257–274.

Bille M and Sørensen TF (2007) An Anthropology of Luminosity. The agency of light. *Journal of Material Culture* 12(3): 263–284.

Bille M (2020) *Homely Atmospheres and Lighting Technologies in Denmark: Living with Light.* London: Routledge.

Bille M, Bjerregaard P and Sørensen T (2015) Staging atmospheres: Materiality, culture, and the texture of the in-between. *Emotion, Space and Society* 15: 31–38.

Bille M and Simonsen K (2021) Atmospheric practices: On affecting and being affected. *Space and Culture* 24(2): 295–309.

Bollnow OF (2011 [1963]) *Human Space.* London: Hyphen Press.

Böhme G (1995) *Atmosphäre: Essays zur neuen ästhetik.* Frankfurt AM: Suhrkamp.

Böhme G (2001) *Aisthetik: Vorlesungen über Ästhetik als allgemeine Wahrnehmungslehre.* Munich: Wilhelm Fink.

Böhme G (2014) Urban architecture: Charting new directions for architecture and urban planning. In: Borch C (ed) *Architectural Atmospheres.* Basel: Birkhäuser, pp. 42–59.

Böhme G (2017) *The Aesthetics of Atmospheres.* London: Routledge.

Borch C (2014) The politics of atmospheres: Architecture, power, and the senses. In: Borch C (ed) *Architectural Atmospheres.* Basel: Birkhäuser, pp. 60–89.

Brennan T (2004) *The Transmission of Affect.* London: Cornell University Press.

Cuff D (1992) *Architecture: The Story of Practice.* Cambridge: MIT Press.

Degen M, Melhuish C and Rose G (2017) Producing place atmospheres digitally: Architecture, digital visualisation practices and the experience economy. *Journal of Consumer Culture* 17(1): 3–24.

Edensor T (2012) Illuminated atmospheres: Anticipating and reproducing the flow of affective experience in Blackpool. *Environment and Planning D: Society and Space* 30(6): 1103–1122.

Edensor T (2015) Producing atmospheres at the match: Fan cultures, commercialization and mood management in English football. *Emotion Space and Society* 15: 82–89.

Edensor T (2016) Aurora landscapes: Affective atmospheres of light and dark. In: Benediktsson K and Katrín AL (eds) *Conversations with Landscape.* London: Routledge, pp. 241–254.

Gehl J (1987) *Life Between Buildings: Using Public Space.* New York: Van Nostrand Reinhold.

Glaser B and Strauss A (1967). *Discovery of Grounded Theory.* Chicago: Aldine.

Griffero T (2014 [2010]) (de Sanctis S, transl.). *Atmospheres: Aesthetics of Emotional Spaces.* Farnham: Ashgate.

Griffero T (2019) Pathicity: Experiencing the world in an atmospheric way. *Open Philosophy* 2(1): 414–427.

Griffero T (2021) Urban atmospheres and felt bodily resonances. In: Catucci S and De Matteis F (eds) *The Affective City: Spaces, Atmospheres and Practices in Changing Urban Territories.* Syracuse: LetteraVentidue Edizioni.

Hasse J (2014) *Was Räume mit uns machen – und wir mit ihnen. Kritische Phänomenologie des Raumes.* Munich: Karl Alber.

Heidegger M (2000 [1919]). *Towards the Definition of Philosophy.* London: The Athlone Press.

Houdart S and Chihiro M (2009) *Kuma Kengo: An Unconventional Monograph.* Paris: Éditions Donner Lieu.

Kazig R (2008) Typische Atmosphären städtischer Plätze. Auf dem Weg zu einer anwendungsorientierten Atmosphärenforschung. *Die alte Stadt* 35(2): 147–160.

Klages L (1974) *Sämtliche Werke, 3 Philosophische Schriften.* Bonn: Bouvier.

Light A (2015) Troubling futures: Can participatory design research provide a constitutive anthropology for the 21st century? *Interaction Design and Architecture(s) Journal* 26: 81–94.

McCormack DP (2014) *Atmospheric Things.* New York: Duke University Press.

Melhuish C, Degen M and Rose G (2016) "The real modernity that is here": Understanding the role of digital visualisations in the production of a new urban imaginary at Msheireb Downtown, Doha. *City & Society* 28(2): 222–245.

Pallasmaa J (2014) Space, place, and atmosphere: Peripheral perception in existential experience. In: Borch C (ed) *Architectural Atmospheres.* Basel: Birkhäuser, pp. 19–41.

Pile S (2010) Emotions and affect in recent human geography. *Transactions of the Institute of British Geographers* 35(1): 5–20.

Rauh A (2018) *Concerning Astonishing Atmospheres: Aisthesis, Aura, and Atmospheric Portfolio.* Mimesis International. https://mimesisinternational.com/concerning-astonishing-atmospheres-aisthesis-aura-and-atmospheric-portfolio/

Riedel F (2019) Atmosphere. In: Slaby J and von Scheve C (eds) *Affective Societies: Key Concepts.* New York: Routledge.

Rose G, Degen M and Melhuish C (2016) Dimming the scintillating glow of unwork: looking at digital visualisations of urban redevelopment projects. In: Jordan S and Lindner C (eds) *Cities Interrupted: Visual Culture and Urban Space.* London: Bloomsbury.

Samimian-Darash L and Rabinow P. (eds) (2015) *Modes of Uncertainty: Anthropological Cases.* Chicago: University of Chicago Press.

Schroer SS and Schmitt SB (eds) (2018) *Exploring Atmospheres Ethnographically.* London: Routledge.

Schmitz H (1978) *System der Philosophie, III. 5. Die Wahrnehmung.* Bonn: Bouvier.

Schmitz H (1996) *Husserl und Heidegger.* Bonn: Bouvier Verlag.

Schmitz H (2009) *Der Leib, der Raum und die Gefühle.* Bielefeld: Sirius Edition.

Schmitz H (2014) *Atmosphären.* Freiburg: Karl Alber Verlag.

Schön DA (1979) *The Reflective Practitioner.* New York: Basic Books.

Schön DA (1985) *The Design Studio: An Exploration of Its Traditions and Potentials.* London: RIBA.

Schön, DA (1987) *Educating the Reflective Practicioner: Towards a New Design for Teaching in the Professions.* San Francisco: Jossey-Bass.

Schwabe S (2021) Order and atmospheric memory: Cleaning up the past, designing the future. *City & Society* 33(1): 40–58.

Sejr K (ed) (2017) *Architecture Creates Value: Buildings, Urban Spaces and Landscapes Generating Social, Economic and Environmental Value.* Copenhagen: Danish Association of Architectural Firms.

Seyfert R (2012) Beyond personal feelings and collective emotions: Toward a theory of social affect. *Theory, Culture & Society* 29(6): 27–46.

Simon HA (1968) *The Sciences of the Artificial.* Cambridge: MIT Press.

Smith RC, Vangkilde KT, Otto T, Kjaersgaard MG, Halse J and Binder T (eds) (2016) *Design Anthropological Futures.* London: Bloomsbury Publishing.

Stenslund A (2014) Kom Nærmere-Lugten i Galleriet. *Tidsskriftet Antropologi* (69), doi: https://doi.org/10.7146/ta.v0i69.27289

Stenslund A (2015) A whiff of nothing: The atmospheric absence of smell. *The Senses and Society* 10(3): 341–360.

Stenslund A (2017) The harsh smell of scentless art: On the synaesthetic gesture of hospital atmosphere. In: Schroer SS and Schmitt SB (eds) *Exploring Atmospheres Ethnographically*. London: Routledge, pp. 153–171.

Stenslund A (2021). Collaging Atmosphere: Exploring the architectural touch of the eye. *Environment and Planning B: Urban Analytics and City Science*, doi: 10.1177/2399808320986559

Stenslund A and Bille, M (2022) Rendering Atmosphere: Exploring the Creative Glue in an Urban Design Studio. In: Stender M, Bech-Danielsen C, and Hagen AL (eds) *Architectural Anthropology: Exploring Lived Space*. London: Routledge.

Sumartojo S and Pink S (2018) *Atmospheres and the Experiential World: Theory and Methods*. London: Routledge.

Sumartojo S, Edensor T and Pink S (2019) Atmospheres in urban light. *Ambiances: International Journal of Sensory Environment, Architecture and Urban Space* 5, doi: https://doi.org/10.4000/ambiances.2586

Tellenbach H (1968) *Geschmack und Atmosphäre. Medien Menschlichen Elementarkontaktes*. Salzburg: Otto Müller Verlag.

Thibaud JP (2015) The backstage of urban ambiances: When atmospheres pervade everyday experience. *Emotion, Space and Society* 15: 39–46.

Tidwell P, Pallasmaa J, Böhme G, Griffero T and Thibaud JP (2014) *Architecture and Atmosphere*. Helsinki: Tapio Wirkkala Rut Bryk Foundation.

Weidinger J (2018) *Designing Atmospheres*. Berlin: Technische Universität Berlin.

Wigley M (1998) The architecture of atmosphere. *Constructing Atmospheres Daidalos* 68: 18–27.

Yaneva A (2008) How buildings 'surprise'. *Science & Technology Studies* 21(1): 8–28.

Yaneva A (2009) *The Making of a Building: A Pragmatist Approach to Architecture*. Bern: Peter Lang.

Yarrow T (2019) *Architects: Portraits of a Practice*. London: Cornell University Press.

Zumthor P (2008) *Atmospheres*. Basel: Birkhäuser.

2 INVESTIGATING ATMOSPHERES
METHODS AND METHODOLOGY

Atmosphere in Urban Design – a Workplace Ethnography emerges from nine consecutive months of fieldwork within the office space of SLA. The nine months were followed by a continued open dialogue until the publication of this book during which additional information, particularly in the form of confirmations, clarifications and permissions to use images, could be obtained from SLA. This chapter contemplates the methodological implications of doing workplace ethnography *in* atmospheres, and, together with one's informants, thinking and talking *about* atmospheres while acting *through* them. This take on investigating atmosphere *in*, *about* and *through* is partly inspired by Sumartojo and Pink (2019), who offer beneficial advice on methods for ethnography on atmosphere. As they say:

> If knowing *in* locates the research and the researcher in the ongoing and emergent flow of atmosphere and knowing *about* is a retrospective look at what it felt like that interrogates a particular instance of configuration, then knowing *through* is a way of using the concept of atmosphere as a route to understanding something else.
>
> (2019: 44)

In this chapter, the *in*, *about* and *through* is paired with a *what*, *how* and *so*. First, *in what* was I as an ethnographer located? Together with whom and with what intentions? Second, *how* did we think and talk *about* atmospheres? How did it feel to go through the enquiry process, and how were methods geared to pursuing the subject adjusted to opportunities and restrictions? And third, acting *through* atmosphere in the enquiry process comes to define data and *so* these data suggest the themes of relevance to the analysis that finally came to shape each of the chapters in this book. Hence, chapter by chapter, the reader will come to understand that atmosphere, as I suggest, involves designers' ways of experiencing, resonating, engaging, understanding, creating, collaging, rendering, anticipating and mimicking quality in experience mediated by their design. These telling present participles formed from verbs have come to structure discussions and analyses within the book and thus some of them are included in the chapter headings.

DOI: 10.4324/9781003279846-2

They serve to emphasise the advantage of expanding enquiries into urban design and architecture from examining plain practice – however valuable it may be – to also paying attention to the atmospheric quality of these practices

By questioning *in what* kind of data the discussions in this book are rooted, below I situate the fieldwork and my entrance into SLA's office space.

Situating the workplace ethnography

Ethnography is a field, commonly used in anthropology, that today facilitates a wide range of social sciences. The situated nature of my authorship is that I am a sociologist by training – a sociologist who has worked within interdisciplinary fields ever since obtaining her PhD. Many of the substantial features of the workplace ethnography consist of what I would refer to as participant observation (Bryman, 2001: 290) throughout my years as a sociology student. However, sensory studies developed, practised, examined and discussed within anthropology (Grasseni, 2004; Howes, 2003, 2005; Ingold, 2013; Pink, 2009), sociology (Simmel, 1997 [1908]), ethnography (Stoller, 1997), geography (Rodaway, 1994) and history (Classen, 1993) have helped shape the empirical studies that I do, and certainly the enquiries into atmosphere in urban design presented in this book profit from my interdisciplinary approach.

During each day of fieldwork, I would participate as an observer in SLA's internal and external affairs. I joined employees at meetings either in-house or elsewhere with collaborators, municipalities or contractors. Together with the designers I went on site supervisions (Images 2.1 and 2.2) that offered a unique chance of doing walk-along interviews (Lee and Ingold, 2006; Pink, 2008). Walking and talking further enabled us – my informants and myself – to share vision (or limited vision, because sometimes we went out in darkness at night) while we heard, smelled and felt the kinaesthetic sensation of rhythm from our gait. Occasionally, on site visits, I asked for the designers' re-enactments (Pink and Leder Mackley, 2014). That is, I asked if they would re-do observations of areas around the city that had informed their drawings, and I would lend them my camera and have them photograph what they once photographed with their own cameras (usually their phones as in Image 2.2) in order for me to see what *they* saw: what they homed in on, how they moved around, what they listened to – in short, what caught their attention and what did not.

With a few exceptions, fieldnotes were roughly written and then rewritten in detail on the same or the following day. While typewritten notes have been useful for systematic searches in Nvivo, my sketchbooks with densely handwritten texts and drawings on blank paper have still become my three-volume Bible. The sketchbooks followed me everywhere, and during interviews, for example, whenever I saw an occasion, I would invite informants to draw and explain in my books rather than theirs (see Images 2.3 and 2.4).

Image 2.1 SLA manager descending a dented cement path during a walk-along
interview and site supervision of CopenHill – a rooftop ski slope.

Photo: Anette Stenslund

Image 2.2 SLA manager photographing CopenHill on site supervision.

Photo: Anette Stenslund

Image 2.3 Employee drawing and explaining during an interview in the SLA office.

Photo: Anette Stenslund

Image 2.4 Stig L. Andersson's drawings and notes during an interview break in the SLA studio.

Photo: Anette Stenslund

'Architects are quite visual', their anthropologist once reminded me, so of course I adapted conversations so that together we could observe and discuss pictures and draw as we spoke. Hence, I consciously probed a multisensory approach to both participant observations and interview techniques that are described in further detail below. Participant observations and informal conversations are at least as significant to the book's discussions as are the 32 semi-structured interviews that were conducted, transcribed and ana-lysed. The majority of the interviews have been conducted in Danish and subsequently translated into English.

In *principle* (and I will return to what this principle implies), I was unre-strictedly welcome to join the working day at SLA and I was treated like any other employee within the two floors of open-plan office. Having read about Yaneva's (2009) fieldwork with Rem Koolhaas, where no employee would have their own workstation, I had prepared myself to face the same kinds of challenges. Much to my delight, however, SLA offered me a desk, a PC and a freshly created email account consisting of my initials (ast@sla. dk). From this base I could access – go to and from – activities in the office (see Image 2.5), keep updated on news, be copied in on email correspond-ence with clients and collaborators, and clients and external collaborators were able to contact me in case of any queries – my portrait photograph was uploaded among those of the staff on their homepage. From my desk I could actively listen each day to the hum of voices within the open office space

Image 2.5 SLA's open-office floor on a regular working day.

Photo: Anette Stenslund

and I was provided with a small refuge from where I could write some of my field notes while they were fresh in my mind and while I was still showing a physical presence. I was told to help myself in the kitchen to warm and cold drinks and snacks, and I was invited to participate in the lunch break – a staff break that probably any workplace ethnographer will know as a rather intense timeslot full of potential for establishing and building new connections with informants, allowing access to informal conversations and providing the occasion for setting up new appointments with employees.

On 6 February 2019, the first day of the fieldwork period, a message was sent out to staff at the Copenhagen office welcoming my presence over the coming year. I was presented in that email as a researcher from the university 'investigating how we work with atmospheres/moods in our projects'. I would elaborate in another email and in the subject line I would describe myself as their new house sociologist, which among the employees came to act as a useful conversation starter. My wording was carefully and well considered. For me, a down-to-earth introduction should serve the purpose of establishing a trusting relationship as quickly as possible with as many people as possible, and one way would be to remove the label of a researcher so that I was more easily considered 'one of them' while still, of course, being different. 'You will be the person with the highest educational qualification in our company', Trine, a team leader, had mentioned to me in advance. 'Well, except for Stig, who is also a professor', she added. I candidly joined SLA as a sociologist and an academic scholar, but clearly, I carried a hope to be thought of as *their* sociologist before my nine-month stay was over. With this motive in mind, I called myself their house sociologist, for when 'house' is used as a qualifier, it often serves the function of domesticating the noun, enveloping it in a homely atmosphere, eliminating distance, enhancing personal relations just enough to make people feel safe and curious, 'but not so much so that they forget there is something strange and special to the encounter' (Whiteley et al., 2017: 225). A house sociologist, so my idea went, would serve rather like the Goldilocks principle – the notion of lying between two extremes of a continuum in a 'just right' position, on a medium domestic scale.

Overtly I confirmed my presence as an academic scholar within the studio, but additionally I was quite concerned to make any employee understand that I had come to learn from their work; that they were the experts and that I was not questioning their skills at any point. I sought to make them comfortable with my presence, stressing that I was not interested or responsible for their HR management as some would otherwise be expecting. Rather, I had come to them with great amounts of admiration and whatever they were willing to share with me about their design practice on an everyday basis, I would treat with the greatest care and respect. Hence, in various ways and varying my choice of words I sought to explain how I as a sociologist would be harmless to their individual careers, and that I would take the role of an outsider to learn about the insider culture of SLA. Of interest to me was how their interdisciplinary team of urban designers would think, speak, act, listen, move around and thus know about, make sense and feel atmosphere in processes of design. Finally, I stressed my particular task of always safeguarding any person's right to anonymity. As mentioned in Chapter 1, with the exception of Stig L. Andersson, all names appearing in this book are fictitious. Moreover, gender has been unsystematically changed to ensure that nobody can be tracked down or identified. The book thus offers no feminist flavour of the topics discussed, which instead can be found in Flora Samuel's book *Why Architects Matter* (2018).

How did we go about communicating?

I interviewed employees with various educational backgrounds and experience, among them interns, graphic visualisers, architects, landscape architects, horticultural scientists, a biologist, an anthropologist, a communication officer, new recruits, long-term employees, team managers and directors. With a few exceptions, the interviews were held in one of SLA's four meeting rooms and would last for about one and a half to two hours. A few interviews were conducted during site supervisions, and a few post-fieldwork interviews took place via Teams and Zoom due to the constraints that were in place by then as a consequence of the pandemic. It was a great advantage to the research project that SLA gently welcomed my presence by budgeting for the hours I would take from each of the employees with whom I conversed. This procedure of allotting the hours for informants was to ease recruitment for interviews and other queries.

But what does an ethnographer talk about with interviewees if the topic of interest is, as some scholars suggest, an example of 'non-representational phenomena' (Anderson and Ash, 2015) not 'easily named' (Schmitt, 2018: 91), and thus better not put 'in the crosshairs' of investigation (Sumartojo and Pink, 2019: 51)? Atmospheres can be approached ethnographically from various angles inspired by phenomenology, affect studies, ANT and non-representational theory, each accentuating differently atmospheres of place, situation, production, co-production, etc. (Anderson, 2009; Bille, 2020; Edensor, 2012, 2015, 2016; Edensor and Sumartojo, 2015; Pink and Leder Mackley, 2016; Schroer and Schmitt, 2018; Schwabe, 2021; Stenslund, 2015; Sumartojo et al., 2019). This book's approach is phenomenological and multisensory: phenomenological because it is interested in the experiences of the informants – their 'sensorial and emotive' way of designing (Pink, 2009: 83) – and multi-sensory because it does not expect its interviewees to 'sit still and talk' (Pink, 2009: 86). None of the interviews carried out for this book are just about talking. All interviews offer opportunities to observe body language, gesture and voicing (see, for instance, Image 2.6). In addition, informants were usually invited to draw with pen on paper whenever it felt natural, and in many cases, we would choose to study pictures and drawings together from which conversations would then emerge. Some meeting rooms could technically facilitate wide-screen access to the archives in order to pursue the images that informants would often refer to of their own choosing in order to exemplify the topics that were on the table.

Image 2.6 Two employees discussing sketches for The Tunnel Factory during an internal meeting in the SLA studio.

Photo: Anette Stenslund

Although no elicitation interviews were conducted (Pink, 2009: 93) – it was *their* and not my documents and images that became the subjects of conversation – our conversations would still circle around and be kicked off by imagery of the informants' own choosing, and thus they played a crucial role for insights that were developed during the fieldwork period. Not only would our document and image reviews help evoke memories and remind us of knowledge that might otherwise have been inaccessible, but with the help of architects and designers I also learned to look at images in certain ways. Hence, they paved the way for a learning process that I had not been able to achieve on my own. On numerous occasions I could not 'read' images correctly until I cross-checked and reviewed them with the designers. Again and again, I would not be able to see what the designers saw, and even in situations where they had described to me what they saw as, for example, errors or successes, I would often not be able to see this before they carefully reviewed the images with me and could point out what exactly to look at. Accordingly, in interviews and conversations we seldom simply talked, but rather we saw together; we pointed, made comparisons, drew and told each other things, and what I later came to write about was confirmed by the designers.

Despite the often vague and diffuse use of the term 'atmosphere', I did not mind explicitly asking about atmosphere, as some ethnographers have otherwise been more reluctant to do (Anderson and Ash, 2015; Sumartojo and Pink, 2019: 51). I would soon learn that what I understood as atmosphere was mostly discussed as 'the felt qualities' of the design, also referred to as the 'amenity value', that would reach beyond rational explanation and be complementary to the 'measurable' parts of the design (Andersson, 2018: 118). But since I am not working with an objective, true and static account of atmosphere that informants could be more or less in line with, the designers' definition was not the most important one to me. If it is true that the phenomenon of atmosphere is far richer than what words can describe, and if I am right to assume that atmosphere is neither objective nor subjective, but a quasi-objective and quasi-subjective phenomenon, then it becomes methodically important to find out how attention is directed exactly towards the situations in which both the object and the subject – in this case the design and the designer – step out of themselves and meet. This meeting is what I am looking into as the 'object' of investigation. I have a trained eye not for material objects' ecstasies or subjects' ecstasies but for ecstatic encounters. I consider ecstatic encounters as atmospheres, and those encounters happen whether the informants explicitly talk of the concept of atmosphere or not.

Stig L. Andersson is an intellectual who has written texts on the subject of aesthetics, including atmosphere, that are a delight to read (Andersson, 2014, 2018). In his catalogue for the Danish pavilion at the 14th international architecture exhibition La Biennale di Venezia 2014, atmosphere is addressed in terms of its dual nature as both a physical condition and an experience that touches our being. Andersson writes (2014: 51):

> Atmosphere is a thin film of enclosure around the world. Without our vaporous, water-filled atmosphere, life on Earth – or indeed, life anywhere – would not exist. But atmosphere is also what you sense in a particular situation. [...]
>
> I believe that it is in the complementarity between the quantifiable facts and the qualitative amenities of a given space that is the secret to achieve the full understanding of atmosphere. As the essential part of life: Physically, sensuously, and aesthetically.

In SLA's archives I find examples of how this dual conceptualisation is conveyed in practical working papers, for example in a guide to lighting entitled *Aesthetic Lighting* (Fartadi-Scurtu and Galiana, 2018). Here, the dual occurrence of the atmosphere is repeated as both a physical reality (as evaporated water that envelops the globe) and as an experienced quality where light is said to be 'atmosphere-creating' (Fartadi-Scurtu and Galiana, 2018: 3). As the guide says: 'proper lighting can instil a certain atmosphere [...] appealing to all our senses. For example [...] lighting manages to add suspense to the entire scene' (Fartadi-Scurtu and Galiana, 2018: 9). The espoused atmosphere is interesting, but it only stimulates the investigations of the current study to the extent that they can shed new light on the enacted atmospheres. I am thus motivated by creating a perspective concerned with how employees might feel and make sense of atmosphere (in the double sense of the word). 'You will always be able to find the answer to your questions in Stig's lyrics', employees often told me when I asked about their design practice – 'but they are difficult to understand', a few would occasionally warn me. I took this warning as an indication that the designers 'on the floor' found themselves on a different plane; not with heads in the clouds but with hands-on experience. These kinds of experiences are therefore pursued in this book, and when I communicated with the designers about atmosphere it usually happened in the sense of a touching sensation of the design, what they in Danish would address as '*mærkbar*', which can mean both a noticeable, perceptible, appreciable and, at the same time, a touchable, tangible, moving quality of their design that appeals to the emotions.

How acting *through* atmosphere together with SLA came to define the data of the research project and *so* to suggest the relevance of themes for its analysis is discussed throughout the rest of this chapter. I will address my ethnographic embeddedness in the study subject, the implications of my physical presence and location in the office, the paradox experienced by being limited by unlimited access, and finally I touch on ethics and the value of not terminating the ethnographical enquiry before findings are discussed with the companies with which a workplace ethnographer engages.

Embeddedness

Atmospheres are ubiquitous, they transcend all boundaries, bodily, locally and globally, and therefore they can never be studied from the outside. We are always *in* atmosphere and atmosphere is *in* us with no exception for the ethnographer. Hence, I agree with accounts that emphasise the importance of the ethnographer openly addressing their embeddedness and recognise how 'this impacts on what we can ever possibly know in such environments' (Sumartojo and Pink, 2019: 37). In the following I will describe how my embeddedness as an ethnographer shaped the study. Did I create what I wanted to see? What made me see what, and what could cause me to see through other things?

Sumartojo and Pink are right in their discovery, I believe, that 'researching atmospheres always involves an autoethnographic engagement' (2019: 39). The question is only how much the ethnographer makes it the focal point of the analysis itself. Edensor (2017, 2013) gives examples that make a virtue out of the autoethnographic engagement. In the present enquiry, however, my own engagement unavoidably conditions the knowledge produced, but no active effort is made to make the study rest on my own experiences as ethnographer. Rather, I consider it my responsibility to take into account that atmosphere is always part of the person or people who act(s) through it, and therefore it is incumbent on me to make clear to the reader from what perspectives atmospheres are experienced. That is, which people experience the atmospheres and where and when. In the next section I will exemplify how I clearly came to shape the study by stimulating the SLA employee's awareness of atmosphere.

Location

My spot in the office determined what I saw. Literally, where I was sitting had an impact on what caught my interest. Evidently, being offered a desk, PC, email and an account providing access to internal archives is an invaluable gift that contributes to the quality of the research project presented in this book, but being located in a particular spot also shapes the study. I started on the 3rd floor and moved only once, after about half a year, to the 4th floor due to 'ordinary rearrangements'. My location in the office has, of course, influenced what I have seen, heard, observed and touched upon with interest. In the beginning I was seated among employees mainly engaged in visualisations, plan drawings, diagrams and 3D and their main contribution was to assist the initial tendering stages. I watched their production of CGIs, listened to their way of talking about them, asked them about their work, heard when they took phone calls, noticed how printouts in a steady stream landed in their wastepaper bins – often I fished out discarded drawings and kept them in a folder (Image 2.7). My location made me quickly draft the two chapters on collaging and rendering respectively (Chapters 4 and 5). Later, I was located with a desk among designers engaged in detail design for construction, which is thematically more prominent in Chapters 6–8.

One thing was that my surroundings including the people inhabiting them made me see certain things; another was that I presumably made those around me see things which they may otherwise not have seen or paid attention to. At one point, a member of staff came to me and said, 'since you arrived, I've started to work with much more focus on atmosphere'. Had I in any way started out with a somewhat naïve ambition to be like a fly on the wall, the field would clearly tell me to forget about that. My role was different from a fly's, and I would become a participant – admittedly not one of the many who could draw, but I could help them present, argue or simply evaluate their performances, which they would sometimes ask me about, especially if I had joined one of their meetings. After I finished the fieldwork and while writing up this book, an SLA director asks the Copenhagen staff

Image 2.7 Random wastepaper bin in the SLA office. The top paper shows a visualisation of trees and is titled 'Atmosphere'.

Photo: Anette Stenslund

to tidy up their desks and shelves a bit early the following day, saying that 'a
little creative working mess' was fine, as photos would be taken 'to show the
atmosphere, the working ambience'. For better or worse self-consciousness
was raising.

For the time being, SLA is undergoing a radical reorganisation which
includes the preparation of a new website for which the requested office
pictures where made. The shakeup of the organisation started after I finished the fieldwork to analyse data. When, after a year, I return with the
first texts and reflections on the project's findings, a manager tells me: 'Your
work helped develop our company. You probably don't know it, but it did'.
Naturally, my 20 years of experience as an ethnographer studying and writing about atmosphere might have helped cultivate my attention to atmosphere, as Schroer and Schmitt (2018: 6) suggest is the case for many scholars
within this field. My trained multisensory attention might have encouraged
and 'given voice' to socially marginalised forms of knowledge which made
it possible to observe, for instance, how CGIs transcend visual communication by involving other senses that are felt atmospherically (Chapter 4).
What I like to add, however, is that whatever this research study might have
contributed, it did not happen due to 'my' merit or 'my' contribution alone:
we made it happen. Ethnographic work is and remains a collaboration, and
the atmospheres described in this book are atmospheres that *we* shared with

each other – employees, clients, external business partners and me as ethnographer. We experienced, consumed, produced and communicated about atmospheres collaboratively.

The paradox of unlimited limits

As mentioned, in *principle* the ethnographic enquiry was given an unrestricted framework, but reality would also reveal constraints along the way and during the fieldwork period. It is important to be transparent about the conditions for the enquiry so that other ethnographers can learn from my experience. I will therefore share a few examples showing indications of a framework being narrowed down paradoxically due to a loose framework rather than intentional control. Some information and knowledge I hold back for ethical reasons.

First, even though I put a lot of effort into winning people's trust, it can still be tricky and rather time-consuming to gain everyone's attention and win their support in a big company like SLA. You are up against a work culture of employees who are used to sticking together in teams around specific projects that normally only include the people who are selected to contribute to a common affair – not to share these affairs plus each person's way of approaching and engaging in the work with an outsider, or to include a strange house sociologist who cannot even draw or produce images, and who is therefore not able to participate like everyone else has been doing. A sociologist doing workplace ethnography will need to enter and breach foreign territory, but how best to prepare the people involved and have them accept one? Although in the previously mentioned email that went out to all staff members it was communicated that it would be of interest not only to the research project and the ethnographer but also to SLA if all employees would engage 'as honestly and openly as possible', several examples were given of employees 'on the shop floor' who either claimed not to have the mandate to speak to me about things that I asked them – it could be about the work process, collaborations in a project or a project's history – or, in interviews, they would return my questions with yet another question: 'What does Stig say?' The repeated examples of people who hesitated to share information with me indicate that we find ourselves in a culture that is trusting in authority and is managed top-down regardless of whether many lead employees showed inclusive attitudes or explicitly stressed to me the horizontal organisational structure of SLA.

It has most probably been a drastic cultural change for some employees not to look after their own affairs only, but instead to share them with me, and I am obviously in debt to the many staff members who embraced my presence as the most natural in the world. Should I have overcome the last hurdle, I am quite convinced that it would have required a renewed top-down effort to gain support. It is, however, not the ethnographer's sole responsibility to secure the management's interest, and the cooperation can be tricky as soon as it involves more people with no time to waste. In the case of the present project, it was Stig L. Andersson who had given me access to SLA's office. He was my gatekeeper. However, I did not know whether Andersson had the other directors' support for my presence. Should I have asked for their approval? I hesitated. I was afraid that if I interfered with the directors' cooperation, it could risk damaging or even ruining the deal I had just landed. As most ethnographers will know, accessing large private companies is no easy task. Instead, yearly in the process I arranged interviews with two of three directors, the third director I had still only met for informal conversations and therefore I proposed giving the three directors a brief, joint update every third month in order to make sure they were kept informed about the project's development. However, my proposal was rejected and I was asked to do meetings alone with a team leader, whom they considered my closest superior, and with Andersson. Work distribution and a limited amount of time and resources is today still my best explanation for this response. I do not believe in any reluctance, yet due to a less clear and strong presence of three directors' management and common support, my access to information was occasionally limited. Partly, employees might have spoken more freely from the heart if they knew 'what Stig had said', and partly, managers might have been careful not to restrict my access to some data if only I had been able to communicate with them how a research project like mine ideally has the potential to surprise with new knowledge; for instance, by putting together reality in new and unforeseen ways. But the prerequisite is of course that one fails to categorise reality before the analysis is done. An example can help illustrate the point.

I won't find out about restrictions until this particular day when I hear that a crucial meeting has been arranged for one of the projects I have set out to follow. The meeting will take place somewhere in the city and one director will participate together with an employee and several external collaborators have been invited. Out of interest, I ask about the meeting. What is it about? What is at stake? What will be discussed? Are there decisions to be made? I am told by one of the directors that the meeting is not relevant to me because 'it's not about design'. As a result, the question arises: what does the director in this case mean by 'design'? What is it that the director thinks is not relevant to their design practice? For me as an ethnographer, all activities leading up to the actual design execution are relevant and even the less expected parts of practitioners' practices might tell me about atmospheres in design. After all, without attending this important meeting, it was not even a given that SLA would be allowed to create their design. The meeting was about politics, economic constraints and negotiations about division of labour and responsibilities. The director's preconceptions about what was relevant and certainly what was not relevant for the ethnographer obviously enabled some studies and closed others. Consequently, in this case I am not able to challenge the director's potential bias towards the separation of politics, economics, and aesthetics. Theoretically, the assumptions can always be challenged, but I am prevented from finding the empirical evidence that will support the argumentation.

Another example of directional instruction I faced on a day while telling one director about some of my observations from external meetings where I had attended employees' visits to other design studios. What struck me was the 'vibe' of diverse meetings and their disclosure of good and not so good collaborations. Some of my observations, however, had been made during meetings for a project that had faced heavy criticism in the public and where SLA was merely in the role as sub-adviser. Even though my story consisted of vivid and thick descriptions full of admiration for the urban designers' skills of balancing the act of making agreements, explaining, defending and insisting on some choices made while also socialising, charming and keeping up a good mood, I was asked not to mention the case by name. So even if my narrative suggested that there is plenty of desirable atmosphere within redevelopment projects that were otherwise maligned for having anything but atmosphere, I was asked to leave the project and help prevent SLA from being associated with the project they were not happy to name. I fully respect their wishes, and it is no surprise that ethnographic works – and especially perhaps workplace ethnographies – are framed. It is however any ethnographer's responsibility to be transparent about how enquiries are framed. Sophie Houdart's ethnography of Kuma Kengo (Houdart and Minato, 2009) was restricted to two selected projects; Albena Yaneva's was restricted to only one project (2009), whereas in this case I was given access to pretty much everything and only along the way I discovered that freedom comes not only with responsibility but also has a limit. I am extremely grateful to have gained as much insight into the company as I did.

Ethics

Being invited to immerse myself in the working day of an interdisciplinary team of urban designers involved a unique opportunity to attend to the sensory and affective experience of the design practice. But as mentioned, with great freedom comes great responsibility in terms of ethics. Hence, with care and consideration I would sometimes withdraw when things got busy or tense at times. In some cases, I double-checked with informants whether they were okay with my presence or how they would like me to behave if we met with collaborators or officials. Sometimes they would like me to introduce myself and the project up front, at other times they preferred to have me shadowing their performance by staying in the background.

After my fieldwork, while analysing it and writing it up I contacted participants to ask whether they still consented to what I was writing, and they were given the opportunity to go through parts of this book before publication. According to Sumartojo and Pink, ethics in atmosphere research also involves 'how our own presence in the field might shape the experiences and feelings of others' (2019: 50). How did they experience having me on board and how was the offboarding? Yarrow (2019) shares an exciting method for eavesdropping on architects' (his informants') self-recorded conversations and responses after his fieldwork ended and after they read his first draft chapters. I am full of admiration for his instructive initiative which I believe can benefit many future ethnographers not solely for ethical reasons but indeed for training and becoming responsive to the office environment's own way of understanding and processing, correcting and taking account of ethnographic research. Findings and drafts for this book were presented to the office not only in written format but also in oral presentations at length for selected employees and as a 15-minute 'teaser' aimed at all staff members on one of their weekly Friday gatherings. The presentations and subsequent conversations were held online and they were recorded and serve as evidence.

Clearly, workplace ethnographies of the kind that I practised should not merely end with analysis and writing; the post-fieldwork conversations are just as crucial as the fieldwork is to the process of recognition. It was the post-fieldwork conversations that informed me about the impact of my research presence in SLA and without these conversations I would never have understood the extent of the impact of the current study. Conversations *after* my presence in the office taught me about the application of the research findings. In presentations I humbly and respectfully sought to tell about what I would see when employees were working with collages, rendering imageries, doing site visits, choosing reference pictures, when they pinned up their narratives on bulletin boards, calculated systems in Rhino, argued convincingly at meetings, etc. and only subsequently would I see how the information was feeding their own arguments as they integrated and sometimes translated my descriptions into their own vocabulary and thus started to reorganise their affairs.

Hence, when the project enabled a focus on what I would come to think of as the urban designers' recognition processes, the designers themselves would start to think in terms of 'methods'. And when I would describe an instance of conceptual ambiguity among staff members, then they themselves would start to think in terms of 'development'. Sumartojo and Pink warn about the research cultures of anthropologists, sociologists and human geographers, all of whom are mainly driven by *understanding* cultures, not being bridged in the future with the research culture of designers, who seek to *apply* and *intervene* in the business practice (Sumartojo and Pink, 2019: 48). In this case, however, I see no definite gap between the research culture that I come from myself and the culture of the urban design studio. In fact, I have witnessed how the 'bridging of cultures' ran deep into a mutual exchange of ideas where as an academic scholar, I would also come to change and learn about the designers' use and implementation of research that is not only 'translated' and 'applied' but built on and converted into something new and different which is worthy of another study. The current investigation thus offers one way of bridging approaches that are normally considered to be wide apart, and the bridging is not taken care of by an ethnographer alone – rather, it *happens* due to the ongoing exchange of ideas between the ethnographer and their colleagues in fieldwork.

References

Anderson B (2009) Affective atmospheres. *Emotion, Space and Society* 2: 77–81.

Andersson SL (2014) Empowerment of aesthetics. *Catalogue for the Danish Pavilion at the 14th International Architecture Exhibition La biennale di Venezia.* Skive: Wunderbuch.

Andersson SL (2018) Atmosphere: A thin film of enclosure. In: Weidinger J (ed) *Designing Atmosphere.* Berlin: Technische Universität.

Anderson B and Ash J (2015) Atmospheric methods. In: Vannini P (ed) *Non-Representational Methodologies: Re-envisioning Research.* Routledge: Abingdon, pp. 37–50.

Bille, M (2020) *Homely Atmospheres and Lighting Technologies in Denmark: Living with Light.* London: Routledge.

Bryman A (2001) *Social Research Methods.* Oxford: Oxford University Press.

Classen C (1993) *Worlds of Sense: Exploring the Senses in History and Across Cultures.* London: Routledge.

Edensor T (2012) Illuminated atmospheres: Anticipating and reproducing the flow of affective experience in Blackpool. *Environment and Planning D: Society and Space* 30(6): 1103–1122.

Edensor T (2013) Reconnecting with darkness: Gloomy landscapes, lightless places. *Social & Cultural Geography* 14(4): 446–465.

Edensor (2015) Producing atmospheres at the match: Fan cultures, commercialisation and mood management in English football. *Emotion, Space and Society* 15: 82–89.

Edensor (2016) Aurora landscapes: Affective atmospheres of light and dark. In: Benediktsson K and Katrín AL (eds) *Conversations with Landscape.* London: Routledge, pp. 241–254.

Edensor T (2017) Seeing with light and landscape: A walk around Stanton Moor. *Landscape Research* 42(6): 616–633.

Edensor T and Sumartojo S (2015) Designing atmospheres: Introduction to special issue. *Visual Communication* 14(3): 251–265. https://doi.org/10.1177/1470357215582305

Fartadi-Scurtu I and Galiana M (2018) Aesthetic lighting. *Principles for Nature-Based Lighting.* Copenhagen: SLA.

Glaser B and Strauss A (1967) *Discovery of Grounded Theory.* Chicago: Aldine.

Grasseni C (2004) Skilled vision: An apprenticeship in breeding aesthetics. *Social Anthropology* 12(1): 41–55.

Houdart S and Minato C (2009) *Kuma Kengo: An Unconventional Monograph.* Paris: Editions Donner Lieu.

Howes D (2003) *Sensual Relations: Engaging the Senses in Culture and Social Theory.* Ann Arbor: University of Michigan.

Howes D (ed) (2005) *Empire of the Senses: The Sensual Culture Reader.* London: Routledge.

Ingold T (2013) *Making: Anthropology, Archaeology, Art and Architecture.* London: Routledge.

Lee J and Ingold T (2006) Fieldwork on foot: Perceiving, routing, socializing. In: Coleman S and Collins P (eds) *Locating the Field: Space, Place and Context in Anthropology*. London: Routledge, pp. 67–85.

Pink S (2008) Re-thinking contemporary activism: From community to emplaced sociality. *Ethnos* 73(2): 163–188.

Pink S (2009) *Doing Sensory Ethnography*. London: SAGE.

Pink S and Leder Mackley K (2014) Re-enactment methodologies for everyday life research: Art therapy insights for video ethnography. *Visual Studies* 29(2): 146–154.

Rodaway P (1994) *Sensuous Geographies: Body, Sense, and Place*. London: Routledge.

Samuels F (2018) *Why Architects Matter: Evidencing and Communicating the Value of Architects*. London: Routledge.

Schmitt SB (2018) Making charismatic ecologies: Aquarium atmospheres. In: Schroer SA and Schmitt SB (eds) *Exploring Atmospheres Ethnographically*. London: Routledge.

Schroer SA and Schmitt SB (eds) (2018) *Exploring Atmospheres Ethnographically*. London: Routledge.

Schwabe S (2021) Order and atmospheric memory: Cleaning up the past, designing the future. *City & Society* 33(3): 40–58.

Simmel G (1997 [1908]) Sociology of the senses. In: Frisby D and Featherstone M (eds) *Simmel on Culture: Selected Writing*. London: SAGE, pp. 109–119.

Stenslund A (2015) A whiff of nothing: The atmospheric absence of smell. *The Senses and Society* 10(3): 341–360.

Stenslund A (2021) Collaging atmosphere: Exploring the architectural touch of the eye. *Environment and Planning B: Urban Analytics and City Science*. https://doi.org/10.1177/2399808320986559

Stenslund A and Bille M (2021) Rendering atmosphere: Exploring the creative glue in an urban design studio. In: Stender M, Bech-Danielson C and Landsverk Hagen A (eds) *Architectural Anthropology*. London: Routledge, pp. 207–223.

Stoller P (1997) *Sensous Scholarship*. Philadelphia: University of Pennsylvania.

Sumartojo S, Edensor T and Pink S (2019) Atmospheres in urban light. *Ambiances: International Journal of Sensory Environment, Architecture and Urban Space* 5, doi: https://doi.org/10.4000/ambiances.2586

Sumartojo S and Pink S (2019) *Atmospheres and the Experiential World: Theory and Methods*. London: Routledge.

Whiteley L, Stenslund A, Arnold K and Söderqvist T (2017) 'The house' as a framing device for public engagement in STEM museums. *Museum and Society* 15(2): 217–235.

Yaneva A (2009) *The Making of a Building: A Pragmatist Approach to Architecture*. Oxford: Peter Lang.

Yarrow T (2019) *Architects. Portraits of a Practice*. London: Cornell University.

3 NATURALISING ATMOSPHERES

THE NORDIC *WABI SABI* WAY

'Do you mow the lawn?', I once asked Stig L. Andersson during an informal conversation about how he managed his own private land in the countryside. 'I don't have a lawn', Andersson responded and perhaps I was right in sensing a whiff of insult because I was not able to figure out that Andersson was obviously not into bourgeois gardening. Andersson, naturally, does a forest floor. This is a private matter, of course, but when I was telling a friend about this book that I was writing, he suddenly interrupted with a curious question: 'Who are they? What kind of people are they? Are they in fact people who love spending time in the countryside? Do they grow things? Are they green-fingered? And what is 'nature' for them?' This chapter is structured around these questions and the characteristics I seek to draw up serve as a backdrop to the remaining chapters.

At SLA they make a point of being a natural kind of people. 'We're not tie types', one employee says casually to his colleagues while they discuss how to prepare for a meeting with clients and stakeholders. Sometimes, in internal emails they call themselves 'nature lovers' and 'tree lovers'. They even published a booklet titled *Why we love trees* (SLA, n.d.). SLA intend a design that fosters and guards 'the grown environment' and allows for its existence on own terms – not cultivated based on beauty ideals and rationales that exist within 'the built environment'. 'Our design process begins with a simple question: what would nature do?' they say on their home page.

In the office, I witness a group of new employees who, during a lunch break, talk about their 'wild' impression of SLA's design style. 'Wild' is their in-situ wording, and they seem to agree about a first impression where the wildness of SLA's city nature 'has no limits'. 'It's like, the wilder, the better', says an intern. These first impressions of a 'wild nature' made by SLA serve as the incentive to, first, consult SLA's well formulated and articulated understanding of nature, and, subsequently, the practised common rules of thumb 'translating' nature ideology into staged nature in urban space. Observations and interviews will demonstrate that SLA's nature-based design practice is aimed at the production of 'nature atmospheres' that resonate with what has been addressed as 'city nature imbrications' and 'wasteland aesthetics'. A more subtle feature of SLA's design approach

DOI: 10.4324/9781003279846-3

makes it almost unavoidable, however, not to think of SLA's business as a kind of Nordic *wabi sabi*.

Theoretically articulated nature

According to Stig L. Andersson, the reason why SLA's preferred design principles may seem wild to some people is because it has *not* been arranged according to modern rational ideals of 'beauty' that cut across wide swathes of art and design history since the Second World War. Rather, SLA subscribes to an aesthetic approach in order to 'solve some of today's hardest urban problems while creating genuine amenity values that, in an unorthodox way, add an extra layer of meaning and quality to the everyday environment' (SLA, n.d.). Andersson defines 'aesthetics' as the complementary opposite to rationality. It involves, as he writes, 'All our senses and all our feelings; that what makes us feel, sense, wonder, discover, think, reflect, imagine and lead us towards new recognitions and new dialogue with each other' (Andersson, 2014: 9). Through his formulation of a theory of aesthetics, Andersson embarks upon a critical discussion of modern society's devotion to rationality, targeting especially architecture's commitment to a strictly logical, mathematical and evidence-based reasoning, structured at the expense of the 'vague, the suspected, the emotional and the aesthetic as ways to reach new insights' (Andersson, 2014: 11). What quantum physics from Niels Bohr has taught him, Andersson announces, is 'that the world in its essence adheres to the concept of complementarity: Everything has two sides. We cannot see them both at once. But the understanding of both is necessary if we are to understand the phenomenon' (Andersson, 2014: 9). Via Asger Jorn's famous 'Silkeborg Interpretation', the complementarity theory is expanded from physics to art and aesthetics, which enables Andersson to point out how architecture too consists of a duality between what he, with G.N. Brandt, calls 'built' and 'grown': construction work at large (buildings and structures) vs. 'nature' (plants, trees, all that grows) (Andersson, 2014: 14).

In an interview I ask Andersson how to understand SLA's 'wild' approach to urban design. Here is his response:

> The wild must be understood as that which is different from the built. Our goal is not to plant wild for the sake of wildness only. Rather, [the goal] is to follow the order of nature. So, for example, one must know what an ecosystem is to be able to put together a plant community; it must be arranged according to the rules of nature. [This means that] one does not plant trees in rows. [When something is arranged in a row, it follows] the rules and logic of the built environment. This is where the material of the grown environment is treated like a building block, and that's not how we do it in SLA. We use the premises given by the grown, and nature does not set up anything in lines or neat circles.

When I talk to employees in the studio about their way of designing, they often start by explaining their basic goal of creating 'nature-feelings' (in Danish, *naturfølelse*) – feelings 'as if you had stood out there in nature', says one. The way to create nature-feelings goes through what SLA officially formulates as their nature-based design approach that recognises what is tied to Asger Jorn's notion of 'nature's order' and G. N. Brandt's distinction between 'built' and 'grown'. Hence, the order that SLA seeks to acknowledge does not rely on carefully thought-through and neatly arranged structures but acts and appears in complementarity to the rationality that prevails within the built environment as untamed, disorderly, unruly, unpredictable and – in some opinions – wild.

There are many ways to understand wildness and nature (Laage-Thomsen and Blok, 2021). During the 19th century green parks were designed in a picturesque style serving a recreational purpose mirrored in understandings of the pastoral landscape deprived of buildings. Here, 'the grown' contrasted 'the built' by offering a refuge from urban living that, at that time, was considered situated far from 'nature'. But 'the grown' did not lack rational order; rather, 'city nature' was cultivated based on a rationale for 'aesthetics of order' (Nevárez, 2007) enforced via high standards of maintenance, surveillance, beautification and decoration (Domene and Saurí, 2007). This is clearly not the aesthetic of nature and the nature-feeling that SLA seeks to design. In what the studio defines as their 'New Nordic Model for city development' (SLA, 2016: 3) they conceptualise a new urban nature, which 'is not just nature in urban areas [...] [or] a greenification of urban spaces' (SLA, 2016: 62); rather, it holds its novelty value by making us 'realise that we are dependent on nature as something other than a resource for material prosperity, and at the same time acknowledge that it isn't dependent on us' (SLA, 2016: 59). The 'new' understanding of nature here seems to equal a kind of inverted pastoral landscape ideology. That is, instead of seeing humanity's all-powerful attitude mastering the growing and green parts of the world, turned on its head, SLA recognises that man is dependent on nature, and the studio sees no dependency of nature 'on us'.

The reason why it is essential to get a solid grasp of SLA's articulated understanding of nature is not only that I cast doubt on its absolute existence, but also that I observe how the enacted nature, designed and exhibited by the employees on a daily basis, differ slightly from the espoused one. Enacted nature is a staged version of socially constructed nature that I have described above, and it does not exist as separate from humans, as if located somewhere 'outside' people existing 'independently' of them, as Andersson puts it. Below, observations suggest how the SLA team of urban designers works with nature as an inherent part of human being. My suggestion is that nature is given as atmosphere by SLA. Hence, SLA's enacted city nature seems to be akin to Böhme's ecological nature aesthetic: one that is given as a bodily and felt state of being – not 'affective', as if people were stroked, hit or caressed by nature's 'effects' coming towards them, but as a state in which clear distinctions between nature and culture collapse. In Böhme's words: 'What we call the environmental problem is primarily a problem of human corporeality' (1995: 14, my translation), and the reason for this is that nature is *in* people as much as people are *in* nature. Nature, thus, is never something to be found at a distance – as something existing somewhere 'behind' us, as if we had lost or dropped nature and now needed to reconnect to it (Böhme, 1989). And with nature within us, it is reasonable to assume that human well-being is an indicator of how well nature is doing and vice versa. Humanity/Nature, they are of one piece.

In fact, it turns out that reality and SLA's eternal development are catching up with this book's findings, because after the book was written in a first draft, and after having shared many of its findings with the SLA team, the studio began a thorough self-scrutiny that reshuffled the company's working procedures. At the end of 2021 they launched a new website that describes the SLA design approach as both *human-centred* and *nature-based*:

> Together we explore, question, prototype, imagine, re-imagine and join our creative forces to create *human-centered* and *nature-based* design solutions.

This marks a significant shift in the espoused practice, which is much closer to the enacted practice that I observe and that I will now turn to.

Nature-based design principles suggesting atmospheres

The following shows in broad terms how employees 'translate' Andersson's theory of complementarity and understanding of nature into a series of easily accessible rules of thumb. Later in the book, especially in Chapters 7 and 8, I go into detail about SLA's design of nature sensations and the processes that lead to their creation. This chapter mainly aims to equip the reader with a general understanding of SLA's nature-based design practice. Before the observed practices are presented, however, there is a need to stress that even if common design principles are identified, they do not diminish SLA's responsiveness to any client's needs and wishes. SLA does not apply principles blindly as if they have the recipe for good design solutions. In many projects, SLA is tied to tasks and predetermined requirements formulated by clients, and with each project, they immerse themselves in the unique 'identities' of places under redevelopment. In parallel, however, SLA rests on a conviction, as explained above. The studio's declared goal is to promote a particular understanding of nature and enable nature experiences in the city accordingly. To achieve this goal, they carry a toolbox containing preferred design principles. These are the principles I am concerned with below.

Big trees, dead trees, many trees for a big nature atmosphere

At SLA they love trees. Many trees, large trees, trees that lean according to sunlight, wind and weather, trees that tremble, trees that die – 'just like in nature', as they say. In their booklet on trees, SLA (n.d.) write: 'In our work, trees are planted for sustainable and functional reasons – but the species are chosen and composed depending on aesthetic quality, sensibility and atmosphere'. Species are chosen depending on atmosphere, and while I am in the studio, I witness how some large trees are imported into Nordic countries from Austria – while others are transported around the globe – to 'add volume to space', as landscape architect Nika says. Large trees are charged with history, and phenomenologically they appeal due to their very size and magnificence, leaving people feeling small and dependent. Moreover, importing large trees – which for SLA often includes exotics (non-indigenous, foreign, introduced plant species) – not only assists by atmospherically solidifying the asymmetrical dependence of man and nature, but also helps reduce much waiting time. If clients and users of the redeveloped spaces in the city had to wait for 'nature' to grow large, 'as nature would have done it' independently, it could take forever before it reached the atmospheric quality desired by the designers. With large trees, however, SLA is beyond that hurdle and can provide urban landscapes that ensure the instant sensation of a 'big' and 'independent nature'. Hence, SLA are seldom into landscape design that involves strategies of complete non-intervention as forms of 'wasteland aesthetics' (Gandy, 2013b) or 'entropy by design' (Gandy, 2013a) within ostensibly empty, abandoned and marginal spaces in the city. SLA's new city nature is, somewhat paradoxically, highly cultivated in order to make its 'independency' felt atmospherically, and perhaps one could even say that the atmospheric quality of 'the grown' is built and thus complementarity dissolves.

Trees are essential to the planet's CO_2 balance, and they can help absorb rainwater and relieve the risk of flooding. Trees can increase biodiversity, strengthen ecosystems, capture airborne pollutants and help regulate temperature. For all these reasons SLA loves trees, but they are not the only reasons. The atmospheric qualities of trees are key to SLA. In fact, they seem to be vital in a way that make up for the costs in terms of CO_2 emissions that the transport of mature species and implementation of exotics might cause. One drawback of importing old trees and exotics, however, is the risk that they die, but at SLA they have learned to embrace impermanence as nature's will. When biotic factors like living plants and trees perish, rot and are devoured, they can offer new habitats for species and finally one day they turn into soil and new life emerges. Thus, dead trees enter the ecosystem. Originally, it was not the intention to place a dead tree in the landscape surrounding Novo Nordisk's new headquarters in the suburb of Copenhagen, but during transport one tree gave up and SLA chose to leave it in the park and embrace its demise (Image 3.1). It turned out that they became quite fond of their decision, and often they mention the dead tree in public presentations as a way of showing their approach to nature. Today other dead trees have found their way into projects by SLA, and clearly dead trees contribute to the atmosphere – we are to feel the 'natural' cycle of life and death we are all part of and over which we have little control.

Image 3.1 Dead tree in the outdoor area of Herlev Hospital.

Photo: SLA

Looking awry, messy, 'blobing' the atmosphere

SLA also likes crooked trees, and at one point they made a convenient arrangement with a nursery that they could have the cast-off specimens that no one else wanted for a token price. 'And the nursery', plant expert Jens tells me, 'coined the term 'SLA trees' to describe all disabled plants'. Today, SLA no longer has this arrangement, because gradually 'other studios have seen how cool it looks, what we are doing, so now they are starting to imitate', says Jens, annoyed that crooked trees have become sought-after. Crooked trees make the urban space exciting and unpredictable, Jens' colleague Magnus, a lead landscape architect, explains to me in an interview, and he adds:

> Also, we never prune their stems, and that gives a completely different feel to the space. You know, pruned trees like the ones you learn to draw as a child. With a straight stem and branches and leaves in a circle at the top. We don't have any of those (see Image 3.2).

Image 3.2 SLA's nature design at Sankt Kjeld's Square and Bryggervangen in Copenhagen. City life (mother with stroller, cyclist and passing car) in the middle of green planting with unpruned trees.

Photo: SLA

On a site supervision, as landscape architect Ole and I walk across a large parking area, he tells me how at SLA they 'go against conventional ways of thinking – like what is architecturally nice-looking', he clarifies as if architecture is conventional *per se*. Ole then points his finger at a plant bed zoning the square:

> As a rule, we need to un-train the construction workers and gardeners we work with, because far too often they place the trees symmetrically or equidistant. That's what they're trained to do, but gradually, as they get to know us, they find out that this is not how we want it.

Ole grins and carries on: 'It should preferably be messed up a little, at least there should be no equal distances, numbers, and so on'. We are on inspection this day to make sure that the square does not become 'too nice', and while I train my eye as a bug detector correcting for 'architectural niceness', 'neat patterns' and 'conventional thinking', I wonder whether this 'messy look' is not just a fashionable hair style but also a trendy urban outdoor design. Ole's colleague Magnus explains to me later in an interview that there's a deeper reason for the style:

> What we do shouldn't necessarily be nice to look at, but the look as such is not really the most important thing. Most important is that it must be possible to be experienced and sensed as moods in urban spaces.

SLA's nature-based design, as described here, acts primarily as an experience sensed mood-wise – what I describe as atmosphere. SLA provides the experience of nature acting outside our control, but in reality imperfection and mess are controlled and cultivated. According to Magnus, 'the look' is subordinate to the mood, but the look, the form and how things are shaped, is clearly not unimportant. As Mark Wigley recalls, 'atmospheric effects cannot be avoided. They permeate architecture' and even the smallest 'choice of representational technique defines atmosphere' in one way or another (1998: 27). Although Stig L. Andersson declares in an interview that he is no longer interested in form precisely because of its subordinate role in comparison to the experienced quality of space, several employees tell me about a preferred form that they tend to use in order to awaken a desired sense of 'nature'. Uffe, one of the lead landscape architects at SLA, says in an interview:

> The sinuous and curved shapes are often reflected in our path systems. They rarely run in stringy lines from A to B. [...] SLA is about creating experiences above all. It's about moods and sensing. This is what makes the whole aesthetic profile and the studio's imprint. And then someone might say, 'well, you just always make those blobs'. Uh, yes. But it works.

I ask Uffe: 'What is a blob?'

'Well, here's one', Uffe says, reaching for my notebook (Image 3.3).

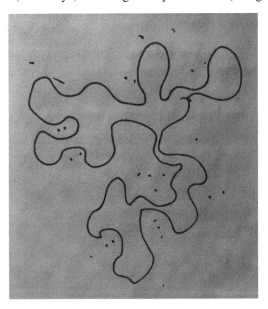

Image 3.3 Sketch by director Uffe in my notebook during an interview. This is a blob: a random, irregular, 'organic' shape of a path system with scattered random dots around it representing trees.

Photo: Anette Stenslund

'There are a lot of SLA projects that look like this, right?' he says while he draws the 'blob'. 'There's a path here, right... And a shape like this, right... and then a few trees all around, right'. Uffe dots the trees and then points at his own drawing as he says:

> That's a blob. When we speak about blobs, this is what we mean and you can see them in many of our projects. That's what Stig became known for. He made these random blob shapes, and the basic principle is that it's a natural form that allows people to move into some spaces where they can feel nature.

Image 3.4 The blob translated into a path system for the Novo Nordisk headquarters.
Photo: SLA

The curved winding paths (Image 3.4) do more than simply ensure human accessibility. The blob has more than a utility function; it offers an added value in terms of atmosphere, not just by breaking with a prevailing grid of straight lines structuring the built environment but by enabling a particular sense of nature – 'It gives you the feeling of being out there in nature', as one architect suggests to me. With Böhme, I would even suggest that what SLA aims at here is not to create nature as if it were 'out there', but that people should feel it – they become that nature in a particular felt way. For instance, when you walk the blob-shaped path system, it will invite you to move in particular ways, you might need to bend your head, move to the side, watch your feet, slow down, because of the way the design behaves *with* you. In this way, the blob turns a passageway into a space that people cannot possibly pass through without having a 'nature-experience'. 'It might seem to be unmanaged – but it's under control. It does something to the atmosphere. It's simply turning the space into another space', Uffe says. This is when 'nature', I suggest, acts as atmosphere.

There are many ways for SLA in general to solicit the expansion of 'nature experiences' in the city.

The rules of thumb mentioned are only a selection of tools that can help represent and stage urban nature, and the degree to which SLA intervenes and how much 'wildness' is tolerated are seldom determined by SLA alone. Rather, nature aesthetics are decided by social collaborations with clients, and I am still to meet a developer who hires urban designers to do 'non design' (Gandy, 2013a: 274) with no or a minimum of intervention. Hence, I see more examples of tamed, staged wildness by SLA, and characteristic is that the design does more than simply 'othering' the built environment. There are indications that the nature-based design acquires a special aesthetic value due to its infiltration into the city. It would probably not maintain the same perceived atmospheric quality if it were designed outside the city, which suggests that it feeds on 'imbricated space', as suggested by sociologist Kevin Loughran (2016). '"Imbricated" suggests the blending or layering [...] of built and natural materials', he writes (2016: 312). Loughran finds a model example of imbricated city-nature in New York's High Line, which is a public park constructed along a disused section of an elevated railroad in Manhattan that contains seemingly 'wild' natural elements such as birch trees and wild grasses that, along with picturesque elements of the built environment – adjacent facades, street view traffic, bits of railroad tracks from the past – form a new aesthetic in the city.

> In these spaces, 'nature' is represented as insurgent – claiming spaces that humans had once conquered. 'City' is represented as decayed – through the rusting and rotting of the built environment.
>
> (Loughran, 2016: 312)

I recognise this kind of nature at SLA. Some collages, for instance, pay tribute to cracks and fissures in the pavement that provide room for weeds – sometimes in wastelands – and generally initiatives arguing for a circular economy to a large extent feed on this kind of city-nature aesthetics that turn the rusting, rustic or worn look of built constructions into something beautiful. The mix of 'built' and 'grown' is prevailing and often staged in playful ways, as when, for instance, a cemented 'blob-like' path system is intentionally pierced to make space for trees to penetrate it (see Image 3.5). In the studio this entangling of built and grown is celebrated as well. One day, while I was in the office, new exhibits were being replaced in the display cases located in the heart of the office space (see Image 3.6). Image 3.7 shows how a root network has formed from the classic SF-stones used for surface

Image 3.5 SLA tree 'in the way' demanding that passers-by watch out if they are not to walk into the tree that is placed in the middle of the path – of course not the exact centre, but slightly to the side.

Photo: SLA

Image 3.6 Display cases in SLA's office. An employee presents grass species in the showcase placed in the foreground. Display case with SF stones and root networks in the background.

Photo: Anette Stenslund

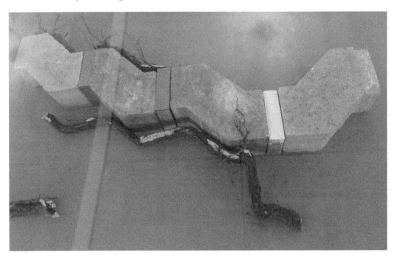

Image 3.7 SF stones and root nets shaped around them.

Photo: Anette Stenslund

paving. SLA therefore does not work 'just' to create more wild planting and less asphalt in the city, but planting is often staged as a force that reclaims space, which in turn awakens an atmosphere that makes people feel it's 'almightiness'.

Nordic *wabi sabi*

When I review all the impressions I have gathered of SLA's design practice during my fieldwork, a kind of dogma begins to accumulate. Of course, there are exceptions to the rules, but some principles boil down to a few sentences, and I find forms of more annotated field notes on the last pages of my notebooks at a time when the fieldwork period is coming to an end. Here is an example:

> Never plant an equal number of trees; be sure to plant *oddly*. Do not buy the straight and symmetrical trees from the nursery; go for the *crooked* and *discarded* ones. All components should be *organised asymmetrically* with *no equal* distance between them. Create *disorder* rather than order – without ending in a mess. There must be some *unpredictability* in the design solution; it should appear *random*, although, of course, it is far from random – *as in nature*.

I did not see it then, but I see it now. The resonance with *wabi sabi* in Japanese philosophy of life is obvious, and I still wonder why no one at SLA mentions it to me. Why don't they say they are making some kind of Nordic *wabi sabi*?

Wabi sabi (侘寂) is an aesthetic that finds beauty in values defined by 'impermanence, humility, asymmetry, and imperfection' as opposed to Western counterparts 'rooted in a Hellenic worldview that values permanence, grandeur, symmetry, and perfection' (Juniper, 2011: 2). *Wabi sabi* permeates all parts of Japanese culture – including art and design. In general, the trained architect Leonard Koren suggests, *wabi sabi* is 'indifferent to conventional good taste. Since we already know what the "correct" design solutions are, wabi sabi thoughtfully offers the "wrong" solutions' (2008: 62). It is pushing it too far to say that SLA advises on 'wrong' solutions, but it nevertheless explains their particular interest in going against conventions and prevailing 'architectural practices' by offering 'unorthodox' ways of approaching urban design. As a result, *wabi sabi* 'often appears odd, misshapen, awkward, or what many people would consider ugly' (Koren, 2008: 62) – just think of the crooked 'disobedient' SLA trees that the nursery would consider of no value.

By mentioning *wabi sabi,* SLA's understanding of nature also suddenly appears in a different light. *Wabi sabi*, Andrew Juniper writes,

> seeks beauty in the truths of the natural world, looking toward nature for its inspiration [...] in order to discover the unadorned truth of nature. Since nature can be defined by its asymmetry and random imperfections, wabi sabi seeks the purity of natural imperfection.
>
> (Juniper, 2011: 2)

Stig L. Andersson is clearly inspired by Japanese aesthetics. He studied Japanese cultural history at the University of Copenhagen in the mid-80s and then went to live in Japan for three years. In an interview, he tells me the following:

> In Japan, I rounded up much of what I had learned here in the North about the importance of materiality, the relationship between *stoflighed* and form, that things are relative, and that there are many truths. Japan made a lot fall into place.

Andersson does not mention anything about *wabi sabi* and he does not elaborate on what he learned about *stoflighed*, form and relativity. He does say, however, as mentioned above, that he is no longer interested in form but rather he is concerned with *stoflighed*. In Chapter 6, I define and translate *stoflighed* as 'materiality', being the atmospheric quality of bodies (human and non-human) as they appear in experience. In the interview mentioned I did not ask Andersson about *wabi sabi* because, at the time, I had only a sparse insight into SLA's practices as summarised above. But curiously, during the nine months I was with SLA, no staff member said a word about *wabi sabi*, and I am therefore left puzzled about the more-or-less direct link that I see. But maybe it is not even that strange, for, as Beth Kempton who worked and lived in Japan for decades reminds us, *wabi sabi* 'is rarely discussed.

Its influence is everywhere, and yet it is nowhere to be seen. People instinctively know what the concept of *wabi sabi* represents, but few can articulate that' (2018: 2). Herein lies the relevance of methodical attention to non-represented aspects of life (Andersson and Ash, 2015).

At SLA they do not talk about *wabi sabi*, but now and then one will hear staff talk about the 'SLA way' of doing things. Employees who carry the 'SLA DNA', as they call it in the office, will know how best to imbue each project that the studio engages in with SLA values. However, there is uncertainty among the employees about who has the DNA and who does not have it. And if you are among those who do not yet have it, how do you get it? And when do you know for sure that you have it? The SLA DNA is not innate, as in biology, but rather suggests it is acquired through experience – as an intuitive form of knowledge, described by Donald Schön (1962). *Wabi sabi* is precisely said to imply an intuitive worldview unlike a logical and rational one (Koren, 2008: 26). Most companies, however, will know that it is almost impossible to achieve a goal if you are not able to formulate clear ideas about where to go. But if the goal is precisely motivated by courage and willingness to dare to engage in uncertainty in order to gain new recognition, to be guided by intuition rather than standard theory, then it may prove sufficient to have a series of rules of thumb that are elastic enough to be tailored to each project.

Perhaps a not insignificant point is also, as Argyris and Schön (1974) have already noted, that architects and designers may not only be challenged to explicate their intuitive and embodied knowledge but also be nagged by a fundamental concern about revealing their know-how. 'Mystique [...] is absolutely central to such professionals. [...] and indeed [they] are worried that too much self-consciousness [...] can destroy the very intuitive skills on which they are convinced their approach to practice is based' (Gutman et al. 2010: 276). In that light, *wabi sabi* seems to be a half-hidden trade secret that goes well with the intuitive work approach. *Wabi sabi* looks for idiosyncratic solutions, one of a kind – not universal, prototype, mass-produced ones (Koren, 2008: 27). This suggests they're more easily enacted than espoused. Stig L. Andersson writes that ways to reach new insights go through '[t]he vague, the suspected, the emotional and the aesthetic' (2014: 11). It is a way of running a studio – the SLA way.

References

Andersson SL (2014) Empowerment of aesthetics. *Catalogue for the Danish Pavilion at the 14th International Architecture Exhibition La Biennale di Venezia 2014.* Skive: Wunderbuch.

Andersson B and Ash J (2015) Atmospheric methods. In: Vannini P (ed) *Non-Representational Methodologies: Re-Envisioning Research.* Abingdon: Routledge, pp. 37–50.

Argyris C and Schön DA (1974) *Theory in Practice: Increasing Professional Effectiveness.* San Francisco: Jossey-bass.

Böhme G (1989) *Für eine ökologische Naturästhetik.* Frankfurt am Main: Suhrkamp.

Böhme G (1995) *Atmosphäre: Essays zur neuen Ästhetik.* Frankfurt am Main: Suhrkamp.

Domene E and Saurí D (2007) Urbanization and class-produced natures: Vegetable gardens in the Barcelona Metropolitan Region. *Geoforum* 38(2): 287–298.

Gandy M (2013a) Entropy by Design: Gilles Clément, Parc Henri Matisse and the limits to avant-garde urbanism. *International Journal of Urban and Regional Research* 37(1): 259–278.

Gandy M (2013b) Marginalia: Aesthetics, Ecology, and Urban Wastelands. *Annals of the Association of American Geographers*, 103(6), 1301–1316: https://doi.org/10.1080/00045608.2013.832105

Gutman R, Cuff D and Bell B (2010) *Architecture from the outside in: Selected essays by Robert Gutman.* New York: Princeton Architectural Press.

Juniper A (2011) *Wabi Sabi: The Japanese Art of Impermanence.* Clarendon: Tuttle Publishing.

Kempton B (2018) *Wabi Sabi: Japanese Wisdom for a Perfectly Imperfect Life.* London: Piatkus.

Koren L (2008) *Wabi-Sabi for Artists, Designers, Poets & Philosophers.* Point Reyes: Imperfect Publishing.

Laage-Thomsen J and Blok A (2021) Varieties of green: On aesthetic contestations over urban sustainability pathways in a Copenhagen community garden. *Environment and Planning E: Nature and Space* 4(2): 275–295.

Loughran K (2016) Imbricated spaces: The High Line, urban parks, and the cultural meaning of city and nature. *Sociological Theory* 34(4): 311–334.

Nevárez J (2007) Central Park, the aesthetics of order and the appearance of looseness. In: Franck K and Stevens Q (eds) *Loose Space: Possibility and Diversity in Urban Life.* London: Routledge, pp. 154–169.

Schön DA (1979) *The Reflective Practitioner.* New York: Basic Books.

SLA (n.d.) *Why We Love Trees.* Copenhagen: SLA.

SLA (2016) *Cities of Nature: A New Nordic Model.* Copenhagen: SLA.

Wigley M (1998) The architecture of atmosphere. *Constructing Atmospheres Daidalos* 68: 18–27.

4   COLLAGING ATMOSPHERES
EPISTEMIC SPATIAL PHYSIOGNOMY

We make conceptual collages to visualize a project's sensuous and emotional character.

—SLA (2021)

In recent decades, research has paid attention to the atmospheric ways computer-generated imagery (CGI) marks the experience of future urban design. What has been addressed in the generic abbreviation CGI has, however, exclusively concerned visualisations that communicate with clients, stakeholders and a public audience beyond designers and architects. This chapter differentiates among the range of CGI used by urban designers. Focusing on collage, which forms one kind of CGI that has received scant attention in scholarly literature, I demonstrate its key function as an epistemological in-house work-in-progress tool that helps designers to refine their vision of atmospheres for future urban spaces. Based on New Aesthetics, collaging atmosphere is characterised by a physiognomic approach to urban space that selectively addresses aesthetic characteristics. Hence, the chapter tackles a discussion that points towards cautious handling of the communicative scope of collages that can be well complemented by other types of CGI before entering a constructive dialogue with clients.

Deciphering CGI

A common way for scholars to debate the recent technological transformation of architectural and design firms has been to study the role of computer-generated imagery (CGI). Particular attention has been paid to the *effect* that CGI may have on urban development (Burrows and Beer, 2013; Dodge et al., 2009; Graham and Marvin, 2001; Kinsley, 2014; Kitchin and Dodge, 2011) and how CGI is part of networks of architects, visualisers, clients, render farm, etc. (Rose et al., 2014; 2016; Yaneva, 2009). Against this background, CGI has been criticised both for its fraudulent alluring properties used for marketing purposes (Jackson and della Dora, 2009; Kaika, 2011) and for debasing the experience of art in architecture: 'Computer-imaging

DOI: 10.4324/9781003279846-4

tends to flatten our magnificent and multi-sensory [...] imagination by turning the design process into a passive visual manipulation', architect Juhani Pallasmaa writes (2005: 12).

There is a tendency to treat CGI in a generic way with reference to the kind of 'visualisation' that many will know from cover pages of magazines, advertisements, posters and websites of developers and architectural firms (Biehl-Missal, 2013; Degen et al., 2017; Melhuish et al., 2016; Rose et al., 2014). Architects call the best of them their 'money shots' – those that serve to persuade and seduce investors and citizens and become front-page news. I refer to these kinds of visualisations intended for pitching a masterplan to an external audience as *renderings*, and they are discussed and analysed in detail in Chapter 5. The reason I ask for greater precision is that lack of specification and differentiation may easily obscure the understanding of what kind of CGI does what, when and how. Technical drawings, plans, sketches, diagrams, renderings and collages are all types of CGI, each making their important contribution to the diversified alignment of specialist fields engaged in visualising urban development. Thus, in order to avoid confusion with the common understanding of visualisation as simply being the act of creating an image, I will henceforth strike a blow for a specific investigation into architectural image making.

Ethnographic work among architects, visualisers and clients has already convincingly argued that CGI in the form of renderings is far from always 'flat' or 'disembodied', as Pallasmaa would have it, but does in many cases evoke 'digital atmospheres' (Degen et al., 2017: 4) or simply atmosphere deriving from the aesthetic, emotional and corporeal impact of the visual encounter (Biehl-Missal, 2013). However, even if the kind of CGI debated in the above works remain composites of many layered elements 'using a wide range of graphic effects' (Degen et al., 2017: 4) processed by several software programs such as AutoCad, Adobe Photoshop and 3D Studio Max (Houdart and Minato, 2009: 88–89), they all belong to the type of CGI that I call 'renderings'. It is no wonder that renderings tend to garner most attention. Due to their extended outreach, they incite the public's interest and therefore also – naturally enough – that of many researchers. In renderings, however, there can be a 'push-and-pull between technical accuracy [...] and atmospheric evocation of what it would "feel" like to be in this new place' depicted by the image (Melhuish et al., 2016: 229). This means that the graphic urban designers who produce the images often find themselves in limbo between, on the one hand, satisfying often competing interests such as demands made on building regulations, financial constraints or requests, clients' expectations or professional artistic interests, and, on the other hand, a wish to devote oneself to the creative, unrestrained design process that ideally allows a design studio to follow its vision, which can in some cases involve the design of atmosphere.

However, since the various types of CGI are produced with different intentions, and since the above-described tension between competing interests is not the reality for *all* kinds of CGI, it is critical to differentiate the field. This chapter will focus on one type of image that in some aspects resembles the rendering, yet in others it differs radically. This is the collage, and both renderings and collages unite under one hat by both being graphic productions aimed at evoking atmosphere. However, with renderings targeted at an external audience, and collages, as we will see further on in this chapter, directed towards in-house collaborators, they also differ in crucial aspects. While today's architectural visual production is often vilified, I seek to re-evaluate the significant contribution of the collage that exceeds the visual communication about a place by evoking an atmospheric and syn-aesthetically felt sense of place.

The computer-generated collage has received little attention in the scholarly literature on processes of designing. From Shields, one will learn that the collage has served as a deliberate artistic method in the architecture of selected renowned architects, showing what she calls a 'collage attitude' or 'collage mindset' (Shields, 2014: 220). Considering, for instance, the marvellous and massive collection of architectural collages made by architect and 'collage artist' Niels-Ole Lund (1990), one will soon recognise collage's ability to stimulate the imagination and offer a critical voice that may manage to transgress conventional architectural practices. But what does the tearing apart and layering of different previously unrelated fragments 'from different levels of reality' (Houdart and Minato, 2009: 86) do to the *process* of developing urban design? It is said that methods involving collaging, montaging and assembling 'convey atmosphere' (Shields, 2014: 114), yet the riddle remains how exactly this happens, and this calls for ethnographically informed knowledge about processes of architectural image making.

After the next section, which offers a conceptual clarification of the collage in urban design, the article is divided into three interrelated parts considering the epistemological, physiognomic and selective ways of collaging, respectively. First, I demonstrate how the collage serves as an epistemological tool in the process of design development and how it deals with uncertainties crucial to the design development. Second, we are to see how physiognomic collaging addresses place identities, and finally I argue how the selective way in which the collage depicts overall atmospheres of urban space not only deliberately overlooks other dimensions of space, but also seems to ignore the outside world's need for answers by speaking exclusively to the connoisseur.

What is a collage and how does it link to the atmosphere of urban spaces?

'Collage' literally means 'glueing' or 'sticking together'. It is a way of making an image in which various cut-out materials – for example: paper, cloth, photographs, or paintings – are stuck onto a plain, two-dimensional surface. Unlike a montage in filmmaking, a collage is a still image, and it has no

3D objects protruding from its surface like, for instance, an artistic assemblage does. A collage stays 'flat' on paper, linen or, more recently and in the case in point, pixels on the computer screen. Only 10–15 years ago, collages made by SLA were created using paper, scissors and a photocopier. Today, however, when designers collage future prospects of the cityscape, it happens primarily on the computer screen and occasionally it will appear on a printout. Collages are typically made in Photoshop – a software program that allows the manipulation of a number of images separately (for example: images of paintings, landscapes, sky, earth, stones or other materials) that can then be layered on top of each other, as if it were a bulletin board with cut-out pieces to be moved around above, below, next to or across each other. In this sense it resembles a photomontage – the composite image of elements from separate sources.

Since Benjamin's Arcades Project, especially, there has been a lively debate on the aesthetic principle of a montage – literary and photographic – with which the designer's collage shares similarities and differences. Both genres still seem to carry the intention, as first formulated by Benjamin, of *showing* rather than *saying* something (Benjamin, cited in Pred, 1995: 11). Whereas drawings 'speak' with collaborators through 'a lot of text and detailed hatchings', Harry, an architect at SLA explains to me, graphic presentations like renderings and collages are made for non-specialist clients in order to 'show' rather than tell. 'This is not to say that drawings cannot be beautifully made and that they lack atmosphere', Harry assures, but for an external audience they are often too difficult to read. Hence, according to Harry, this shift of interlocutor followed by the adjusted choice of CGI – from drawing to presentation – allows an enhanced focus on the atmosphere as the subject of conversation. So even if atmosphere is omnipresent – it incontestably appears everywhere – the *way* it appears depends on the people who attend to it, including their skills and cultural capital, and the designers seem to be aware of this.

Pivoting on an atmosphere approach to the study of collaging urban designers has the advantage of paying particular attention to the fact Harry here draws attention to: that atmospheres are conveyed according to the situated encounter. In this perspective, atmospheres are not said to be *contained* in collages, the collages do not *hold* atmosphere either, atmospheres rather take shape due to the interplay between collages and people who engage with them. With this approach, the distance that might have otherwise occurred between the collage, its maker and given 'recipients' dissolves. Thus, even if I sometimes refer to a 'recipient' or 'viewer', this fictitious person is not a passive receiver of any kind of message sent from designers via a collage, but rather an active co-producer of the atmosphere that is conveyed *within* the encounter. This also means that when I tune in on designers working with atmosphere in the city, it is never claimed that urban designers can wilfully create atmosphere either on-site or in a collage – but they do co-create atmosphere, I suggest, on site and through collaging in the office. I now turn to the epistemological contribution of the collage made possible by its aesthetic, somewhat critical form, asking questions rather than answering them.

Epistemological collaging

Ten to fifteen years ago, SLA could pitch projects using collages. Today, however, collages are only on rare occasions used to market project proposals for urban renewal. Clients have become too expectant in the sense that they wish to see how a project turns out 'in reality', says Yoko, an architect and expert in graphic presentations. Renderings are used for marketing purposes because they can present in almost photographic detail how a project might look after its completion. A collage, in turn, does not necessarily show what a project ends up being like in all its aspects. But what purpose does the collage then serve? I ask Yoko, and she explains:

> When I have a question I want to investigate, I make a collage. And then it'll start asking questions in return that I'll need to respond to by way of describing the collage to my peers. I then try to put some words on the collage […]. Through this process I become more distinct in the way that I express myself about the concept, and I get better at putting words to what we decide to draw.

In most cases the collage seems to bring the expressive capacity of the designer to a higher standard. The way it manages to do so is by having a free and playful form that allows still vague ideas to be sketched out and given a visual expression that can then be tested further in a dialogue with colleagues about how to progress. The collage is not the only visual medium that helps vague ideas become more definite and clearer.

It is difficult to determine exactly when collages are introduced into the design processes. Design processes are non-chronological, iterative and different from project to project, and 'If there *is* a process, it is certainly not linear', Houdart also learns (Houdart and Minato, 2009: 58). They are part of the initial stages, but what comes before and after is different. I have seen how meetings discussing inspiration from previous projects and reference images found on the web or social media can often precede the collages. Frequent printouts of images found through various channels are introduced at meetings where some are discarded, and some are pinned on bulletin boards arranged as mood-boards or, more chronologically, included in overall narratives of given projects. Everything is talked through repeatedly, and at some point, while arguments are hardening, the collage finds its form and enters into the extended sequence of team meetings in order to support the process of recognition. Yoko helps to formulate its epistemological contribution. She talks about how she would normally bring collages to meetings with peers in order to become aware of matters that she has not

Image 4.1 In-house collage by SLA. A layered composition with picturesque glow-
ing sky, harbour quay in front, two people jumping into the harbour
basin on the right. Trees and greenery from the left weave into the paving
in the centre of the image.

Illustration: SLA

been able to consider herself without the collage and the peers. In the following, she clarifies with reference to the collage in Image 4.1:

> I'll present it and say, like, 'in this collage we've examined this meeting
> between nature and existing context. We think it could be exciting to
> work with this edge near the waterfront' or whatever the qualities are
> that I find interesting. So [the collage] is taking an extract of what we
> want to work on, and then the others might challenge the idea: 'Oh,
> shouldn't there be more green?' or whatever. And even if the quantity
> of green was not my intention with this collage, then I nonetheless find
> out something new from the collage. I may not find a solution, but I
> become aware of issues that I need to take into account for the next
> draft that I do.

During the fieldwork, different adjectives are used when employees in
the studio speak about renderings and collages, respectively. Renderings
are predominantly 'smooth', 'polished', 'finished', 'realistic' and carefully
composed, whereas collages are 'rough', sometimes very 'abstract' and
made 'in a jiffy'. Both renderings and collages are often very 'beautiful'.
According to the design specialists that I have spoken to, a good collage can
be made by a single employee in under an hour, whereas the professional
rendering will usually require weeks of preparation with multiple parties

involved. Renderings obviously are far more time-consuming, much more expensive and they are 'smooth', I argue, in the sense that they must show consideration for what SLA employees would often call a 'many-headed client': they would need to assuage potential worries or concerns of all parties, be they citizens, politicians, traders or proprietors, by providing answers, and thus ensure the appeal to as many people as possible in order to market project solutions. In other words, renderings literally need to be appetising and easily digestible to the greatest common denominator, whereas collages can ask questions and show a kind of resistance in this respect – even to its own 'creator'. The collage is of course not exempt from compliance with market requirements, but because it is often kept as a tool for internal use, it can act as a 'work-in-progress image', as Yoko says, and she explicates:

> We've free hands when we work with collages. You don't have to sit and spend a lot of time perfecting it, because it's a working tool to get wiser and learn. It can be very abstract, it doesn't really matter – it's more about the larger whole.

Since collages, unlike renderings, often avoid a direct confrontation with external opposing interests, they are able to fulfil a role as the abrasive and 'rough' sparring partner of designers, thereby providing the design process with intellectual vigour. Reminded by the resemblance to a montage mentioned in the conceptual clarification above, a collage – even if it does not hold the same critical capacity as Benjamin's montage methodology might do – still shows some kind of 'resistance' or 'roughness' that is productive for the design development. Allan Pred (1995) has described the characteristics of a montage as interruption and deconstruction of context, and such criticism is obviously less pronounced in a collage, which is ultimately concerned with marketing coherent and constructive project proposals for urban development. The collage therefore cannot afford to obstruct the context in the same critical way as can the scientific endeavour, but still it holds some degree of resistance. 'It's edgy', Yoko tells me, hinting at a certain irritability.

What is remarkable is how the collage transcends its visual expression in several ways, and how it provides a fairly small refuge for architectural and designerly creation at a distance from outside interference. Both instances seem to reinforce the design production of atmosphere. To begin with the latter, the refuge: there is no guarantee that a collage will promote atmospheric urban design. It is more telling to say that the collage offers a workspace for whatever a group of designers considers crucial for its design before external partners have their say. What different companies might highlight as their strategies varies, and I have interviewed companies that do not make use of collages at all. But in the case of SLA – a studio with a strong appreciation of and admiration for the aesthetic experience and creation of atmosphere, clearly the minor refuge provided by the collage seems to facilitate the 'atmospheric grounding' of projects. Stig L. Andersson explains to me in an interview:

> It is important that everyone who works here is aware that our real task [...] is to create aesthetic value. [...] The means is the utility value of projects, [for example to create] climate adaptation in the urban space, [but] the goal is the atmosphere.

For companies of that type, with a set of basic values prioritising the aesthetics of the urban space, the collage may well serve as a platform paving the way for the atmospheric design, although the 'free' space of the collage is often limited to the sketch proposal phase of a project. Qua the small breathing space offered by the collages, however – existing from a distance to varying external parties who might find peace of mind as soon as they are *certain* about SLA's performance – the collage seems to deal with uncertainties in ways that stimulate the designers' creativity. For example, they examine questions they would not be able to either ask or answer prior to the process. The collages therefore encourage the designers' open attitudes and ability to seize potentials and explore opportunities rather than dismiss them at an early stage.

As for the visual excess of the collage, it seems to surpass its purely visual expression partly by developing designers' language and partly, as we are to see below, by being attached to an embodied multi-sensoriality. The collage develops the designer's expressive skills by initiating discussions among colleagues, which is crucial for the teams' internal recognition process and their ability to develop the projects conceptually during the design process. Hence, the collage helps designers to pin down in words visions of future atmospheres in urban space – atmospheres which are often said to be ontologically diffuse, undefinable or non-representational (Anderson and Ash, 2015; Böhme, 1995: 22; Schmitt, 2018: 91). Beyond the collage's epistemic contribution, it seems important not to neglect the role of vision in the collage. In a branch that has otherwise been severely criticised for its bias towards vision, suppressing other sensory ways of conceiving built and grown environments (Pallasmaa, 2005), we will see below how collaging often exceeds pure vision in order to depict the identity of urban space. I will explore the visual excesses of the collage through what I call physiognomic ways of collaging atmosphere.

Physiognomic collaging

Lily, who had worked intensely with collages throughout the sketch proposal phase for a project that was now approaching its end, showed me how she had sought to 'translate' some of her collages into renderings. 'Look here', she said, handing me some printouts:

I think I managed very well – in this case at least – to have the atmosphere converted. But you see what I mean by the collage capturing the atmosphere more directly, right? Anyhow, renderings tend to appear more far-fetched because we have to take the place into account. In the collage, in turn, you can still feel the roughness.

I understand Lily's statement as an indication that the roughness played a crucial role in evoking the atmosphere of this place. Somewhat paradoxically, however, and even if the rendering had been successful in Lily's opinion, it seemed to lose its atmospheric 'touch' of roughness because of the attention given to 'the place', as she said. With good reason and under different circumstances, one could have assumed that the place is essential to and inseparable from its atmosphere, so what Lily seemed to suggest gave rise to more questions than answers. Did she mean that the collage was completely detached from the place which was under redevelopment? Was the collage generic in the sense that it could have been created for any project? Or would a sensitive attitude towards the place simply happen at the expense of the atmospheric touch of the collage?

The answer was none of those. I have yet to meet an urban designer who would be satisfied with a generic project solution. What normally would mark a 'good project' in SLA – projects that the urban designers would vouch for in public – would be the attentiveness to the unique characteristics of the place under renewal. Showing a sensitive attitude towards the 'place identity', as they call it, helps to develop 'customised' visions for clients and users. Lily herself tells how, in order to produce graphic presentations, she would usually leave the studio for half a day or more to collect 'data' on site. I witness how several of Lily's colleagues do the same thing; they leave their desks more or less systematically, sometimes they investigate sites by using drones (Image 4.2), sometimes small teams leave their desks for half or whole days (Images 4.3 and 4.4), like Lily says she does, and sometimes they might simply pop over to see the landscapes they are developing on when they are on their way to or from the office anyway, early mornings or at night (Image 4.5). Usually they use their phones to memorise how sites look, and they photograph everything from kerbstones, soil or paving, large stones, old rails, reinforced concrete rubble, rubber tyres, masonry or abandoned buildings, plant species, trees – for all of which the designers have a 'trained eye' (cf. Grasseni's 'skilled vision', 2004) to detect and 'translate' into a 'material narrative' to be developed conceptually though the design process.

Even if Lily's claim about her site sensibility would make any ethnographer choke, the studio at least seeks to appear observant of the material aesthetics on site. As one of Lily's colleagues explains when I ask about their procedure for data collection:

Image 4.2 Drone photography of the area around The Tunnel Factory in the North Port of Copenhagen. Labels on the photo bear witness to how many drone photographs are used for recording building materials, discovering historical traces of a former use of the areas and finally for recording plant life.

Photo: SLA

Image 4.3 SLA employees on site visit registering materials and discovering their potential for the conceptualisation of The Tunnel Factory.

Photo: SLA

Image 4.4 SLA employee discovering the area for The Tunnel Factory.

Photo: SLA

Image 4.5 SLA employee in walk-along-interview and re-enactments of an on-site
visit at Copenhagen Village.

Photo: Anette Stenslund

Basically, the way we gather empirical evidence is shaped by the architectural method. So it's photo registration, rather than fieldwork. [...] It's not about talking to people.

[...] If we were to do interviews it would take a lot of hours and that's expensive. [...] It's not that we don't want to. Or, of course it's a pity that people [in-house] don't see the value of talking with locals, but actually what is regrettable, I think, is that clients don't ask for such a service.

There are exceptions where SLA has conducted interviews on a small scale, but basically they acquire their knowledge about sites through printed and online material and through on-site visits with an 'educated attention' (Gibson, 1979: 254) primarily to materials, built and grown, that they see 'carry a narrative'.

Part of urban designers' skill in selecting 'photogenic' materials to be brought into a collage rests on a trained eye for what Böhme calls 'the ecstasies of things' (2001: 131). Böhme refers to his idea about ecstasy in order to demonstrate how things and 'half-things' (e.g. wind, sound, smell, light) within our surroundings are anything but passive entities and occurrences, as they 'extend beyond themselves [...] in complex relational interactions with other entities' (Dorrianin Böhme, 2017: xxi). By definition, what are termed 'half-things' differ from things because of their less stable presence: they seem to come and go and still it makes little sense to ask where they have been in the meantime (Schmitz, 1994: 80). Examples of half-things are wind, sound, smell and light conditions So, for instance, when SLA choose to show in a collage four boys standing in the middle of an open steel structure facing the bay with each of them holding bath towels in the air that are taken by the wind, then the collage communicates 'a meeting with the wind' and not, say, how this place might be particularly good for putting out your washing to dry (Image 4.6). It is the ecstatic way that things and, in the case mentioned, the wind as a half-thing mark how we feel about a place, without us necessarily being able to point out things or half-things as the cause alone for how we feel about a place that the collaging designers pursue.

The way that urban designers produce a collage can be seen as a token of their delicate sense of the ecstasy of things and half-things. But I see their work as even more than that, which is why I make use of a related concept from Böhme to better analyse the collaging of atmosphere. This is the concept of *physiognomy*, of which a reinterpretation can be found in Alexander von Humboldt (1844). Physiognomy was traditionally used to denote a human being's 'inner' character with reference to their 'outer' appearances, but Humboldt both de- and re-constructs this tradition anew, relying in turn on the work of Johann Wolfgang von Goethe, Carl Gustav Carus and Herbert Lehmann. I will here stick to Humboldt, who carries out

Image 4.6 Conceptual collage for the City of Odense.
Illustration: SLA

a physiognomic reading not of humans but of plants and landscapes, which he considers the artistic and graphic representation of nature.

Humboldt argues that the very shape of a plant evokes a distinctive feeling of a place – which Böhme interprets as atmosphere (1999). According to Humboldt, it is the job of the graphic artist (and painter) to suggest this sensory impression of both types of plants and landscapes so that 'their moods' can be experienced by an outsider – that is, people not familiar with the plants and the landscapes represented by graphics. The ambition of such physiognomic landscape communication is, however, not just to pass on individual experiences of graphic artists, but about creating a fundamental cognition or knowledge: 'The subject of physiognomy is not the here and now and the individual experience, but the general, the typical' (Böhme, 1999: 100, my translation). Hence, there is a difference between ordinary descriptions of landscapes and a landscape physiognomy which is about generalisability, and it is the task of the artist to address the typology of landscapes through suggestions that reveal, for example, the Nordic feel of Scandinavia, the Mediterranean touch of the Italian sky, etc. Humboldt writes:

> Who doesn't feel [...] differently tuned in the dark shade of the beech trees, or on hills wreathed with solitary fir trees; or in the grassy field, where the wind whispers in the trembling arbour of the birch? Melancholic, serious, or cheerful images call up these patriotic plant forms that emerge in us.
>
> (Humboldt, 1806: 13–14, my translation)

What counts for Humboldt is the artist's ability to select the constituents of a landscape and find the basic features that in their entirety make it possible to intuitively experience the typical feel of a particular place – or what constitutes its atmosphere. It is natural enough to think of urban designers who produce collages instead of Humboldt's artistic graphics and landscape painters, and it seems to be a logical move to consider the ecstatic radiance of atmosphere from plants and landscapes also to be emitted from the urban spaces under redevelopment. Graphic designers in SLA showed a remarkable trained 'way of sensing' the overall feel of a place due to its visual surfaces. For instance, when they 'photo-registered' a place, they seemed to embark partly upon pre-given constraints concerning utility issues addressed by clients, partly on sustainability goals and standards advised through principles in circular economy embracing recyclability and the reuse of materials (which incidentally fits well with *wabi sabi* aesthetics), and partly on the physiognomic aspect of what would hit a spectator – including the designers themselves – like a *punctum* effect (Barthes, 1981). The *puncta* and the *wabi sabi* aesthetic I analytically see present in SLA's projects so that what hits like a bullet point (•) in experience is often a quality that is unexpected, surprising or – as I suggest in Chapter 7 – astonishing.

Back in the office, the designers work on identifying, naming and categorising the objects photographed, and at some point, the aesthetic potentials and felt expressions that the designers want to extract from the materials are discussed over and over again in order to be discarded or pursued further (Images 4.7, 4.8 and 4.9). The procedure involves some careful choices of image-making based on the atmosphere they seek to support through their design. The urban designers address the constellation of constituents that are telling of 'the feel of the place' in order to make them appear in the collage, suggesting the atmosphere of a place through the extraction of things and half-things. As Yoko said to me, 'it's about the larger whole, not the parts in themselves'. Without her knowing, she would then speak along the lines of Humboldt's 'total impression' of a region (1806: 11).

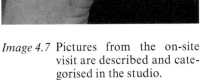

Image 4.7 Pictures from the on-site visit are described and categorised in the studio.

Photo 4.7: SLA

Image 4.8 Categorisation of on-site photos.

Photo 4.8: SLA

Image 4.9 SLA employees preparing a first pitch for the Tunnel Factory based on their on-site visit.

Photo: Anette Stenslund

This way of collaging through a delicate typological understanding of the correlation of felt sensations radiating from a selection of visual surfaces of a place I consider a physiognomic exercise that goes far beyond the intuitive sense of things' ecstasies taken separately. The physiognomic approach to urban space is a 'reading' of material surfaces decentring and 'leaving' the objects like in ecstasy, and that – with the collage – meet, touch and merge with a receptive viewer's sense of a given place under redevelopment. When the ecstasies are selected, taken apart and put together in new constellations tied to place-identity, I suggest we are talking about physiognomy. 'The collage', says Harry, 'serves a bit like a stock cube holding the essence of the distinctive characteristic of a place'. Harry's metaphorical way of describing the collage 'holding the essence of a place' tunes directly into Humboldt's thoughts on landscape physiognomy. To Humboldt, however, essence would not be tied to an inner, hidden or underlying being that can only be addressed hermeneutically. Rather, the essence of a place is a surface character; it can be sensed and felt aesthetically – not as a principle of beauty but as the principle for perception and sense cognition. That is why I tend to consider the physiognomy of the collage as an aesthetic kind of physiognomy.

I began this section on physiognomy with an apparent paradox: the less precisely an image addresses place, the more atmosphere it seems to evoke. As Lily said, she would capture the atmosphere much better through collaging than through her photo-realistic rendering. As I have argued, this does not imply that collaging is detached from the place on site, but rather that the method by which collaging designers make a place appear is very different from the way a rendering is made. I have argued that what designers bring home from their on-site visits in the form of photographs holds the physiognomic potential of things and half-things that can help give others the impression of what it would be like to be *in* a place in terms of a bodily, material and sensory presence, described more thoroughly below.

Selective collaging

In SLA there would be no talk about physiognomy and only occasionally you would hear the employees talk about atmosphere. Instead, more often, words like 'felt' qualities (*mærkbart*) complementary to 'measurable' parameters (*målbart*) would be used in order to describe redevelopment projects. 'The felt is the emotional and sensed', Jonathan from the communication team explains. 'For us', he adds, the felt 'is about adding value and creating quality of life for people'. In the studio, the urban designers would work with 'the felt' when working with the aesthetic experience of atmosphere, while working with 'the measurable' would mean addressing scale, quantity and relational placing of things, trees and buildings, for instance. While technical drawings are particularly good at addressing the measurable aspects of a place, graphic presentations are extraordinarily good at addressing the felt – atmospheric – dimensions. However, as the reader will know by now, there are differences between presentations: where renderings must live up to high expectations attending to almost *any* dimension of a place – measured and felt, the collage will often *selectively* address aspects of a place to be felt. This observation will be expanded upon below.

Although renderings do not show reality but merely an imagined future reality, the expectation is nevertheless that they will match reality and not embellish it; they are produced in order to enter a dialogue with external parties about a future cityscape. It is a different case with collages, where standards of reality might still matter, but they matter differently – that is, atmospherically. This is not to say that collages are never discredited like renderings sometimes are. After all, there are good reasons for SLA to be very secretive about the collages right up to the end of a project, when what is known as a 'workbook' (an internal log containing a summary of methods and general considerations) is occasionally handed over to the client for transparency purposes. One of the most obvious reasons to keep the collages for internal use would be to avoid any misreading. Yoko explains how she sees the dilemma of fact-checking reactions to fictitious imaging:

> There are [people] who are very pragmatic [when presented with an image] and they immediately tend to misunderstand: 'Oh, do you really want a zebra there?' [they might ask] even though the important thing is not the zebra here or there but maybe something completely different […]. What do I know? It might hold an aesthetic value or create an effect or something. If you then start to edit in order to correct the image, the collage loses what I think it can bring forth, namely, to relate more freely to the way of addressing the call.

Collages can be 'very abstract' and 'very personal', Yoko adds, and therefore one should not get hung up on the detail about what is where. Standards of reality are not what matters most in a collage. To exemplify, in a collage (Image 4.10) made for the development of Levantkaj, a place in the North Port (*Nordhavnen*) of Copenhagen, some large heavy rocks are placed in the foreground, the sea is rough and rain is pouring down from a cloud-covered sky. The sea surface seems almost beaten by the rain. In reality, this spot at Levantkaj may not look exactly the way depicted by the collage: the stones might be oversized and they are not to be found exactly that close to the cargo quay; the wind would normally come from a different direction, which makes the scenario in the collage, with waves hitting the dock, a rare sight. But this is not really an issue here.

In the collage the designers have deliberately left out some ontological layers of the place at Levantkaj – the wind speed calculation, for instance, is ignored – and with the collage they allow themselves to focus on some

Image 4.10 In-house collage for Levantkaj in Copenhagen.

Illustration: SLA

single dimensions of the place that express its atmosphere. Hence, what the designers aim for in this specific case of the Levantkaj collage is to evoke the 'roughness' of the place, and the roughness discloses itself when the viewer encounters rocks, wind, rain and sea in the collage. One should therefore no longer consider the 'mere' appearance of the collage to the eye. 'The invisible is the firm component of architecture', as Hasse quotes the contemporary artist Jörg Sasse (Sasse, quoted in Hasse, 2012: 43), suggesting that the collage seems to have an associative surplus that can bring forth 'more' than what meets the eye. For example, the visual encounter might enable a sense of touch, smell or hearing (Morselli, 2019; Stenslund, 2017). In the Levantkaj collage, rather, one should be able to imagine how it would feel to be *in* this place: feel the blowing of the wind, the bite of the cold, the roar of the waves, the play of dim sunbeams far out on the horizon and the grip of the concrete under your feet. It is qualities like these that roughly outline what we may understand as atmosphere: the felt sensation of being *in* a place, with things and half-things that can no longer be differentiated from the person who experiences them – not even if this experience is brought about by a computer-generated image like the collage. As Pallasmaa says – without, however, recognising the value in the digital production – the collage invigorates the experience of tactility (2000: 80). Or, as I argue, the collage transcends the visible through its way of gesturing atmosphere – it points to something other than what is seen by the eye – just like other kinds of images may manage to do. What one sees is therefore not necessarily what one *can* see, but something fairly different that should be picked up from the image by multiple ways of sensing.

It might ease understanding of the selective way that a collage addresses urban space to consider the urban space as analytically divisible into different ontological layers: a mathematical, a symbolic, a social and an embodied space, as Hasse suggests, that can be lumped altogether in a situational space (2014: 21–42). Through such an analytical lens, a rendering will 'speak' to as many layers as possible – it will seek to address the complete situational space: the mathematical space, which can be measured and relationally arranged; the symbolic space, that can be intuitively interpreted according to cultural production of meaning; the social space, that acknowledges how taste, style or activity appeal to different segments within the population; and the embodied sense of a place, which designates felt impressions of a space (Hasse, 2014: 21–39). Unlike renderings, however, the collage clearly does not deal with space in all its detail, but rather quite the opposite: it seeks to evoke an overall impression of a space.

Hence, when issues pertaining to CGI are negotiated concerning, for instance, camera angle, scale or zoom adjustments – downscaling (zooming out) in order to show more of the landscape and surrounding area or conversely upscaling (zooming in) in order to show detailed sectional views (see Chapter 6), this is a matter of engaging in a space in terms of measurement. Calculable solar studies, shadow diagramming, density studies, approximate location of construction work are measurable properties as well, and such ways of approaching a place by scale are very much reflected in renderings but absent in a collage. Moreover, whilst not excluding that collages can communicate through symbols, this is not its supporting language either. Collages are not primarily to be read but sensed and felt. This is due to their alluring atmospheric power that *involves* the eyewitness to a degree that exceeds pure vision. A collage is 'ready to hand' in the sense that it is felt intuitively without taking any notice of intellectual, objective or educated ways of evaluating the architectural design at arm's length. Again, by way of comparison, when urban designers in renderings portray a man with dreadlocks and tattoos or when they present furniture by selected brands, they normally do so deliberately, expecting an allegorical reading by a recipient reacting to a language of symbols, cultural value and taste, thereby addressing what can be defined as the symbolic and social space. However, when the collage seeks to make one feel a place, its aim is slightly different. Suddenly, the atmospheric effect of the design becomes of primary importance: it favours the aesthetics.

Aesthetics account for only part of the complete space ontology, however, and for that very reason, it is prudent practice for an urban design firm to refrain from sharing such a partial piece of the complete design solution with clients until that very moment within the design process when the aesthetics can be complemented with other ontological layers of urban space. This is because it is often not the aesthetics that forms a client's main incentive to seek out a design company for advice, but matters of user behaviour, economic development or environmental issues. Hence, whether clients seek advice to address problems concerning, for instance, rainwater management, CO_2 reduction, noise reduction or crime-fighting efforts, urban design solutions naturally need, first and foremost, to solve such issues on demand with high utility value. In such situations it seems fair to ask what purpose a collage with felt qualities of a space may serve.

In a company like SLA, which rests its vision and practice on the conviction that everything has two halves: a rational and an aesthetic one, a measurable and a felt one, one of direct utility and one of indirect utility (amenity value), the intention is not only to solve a problem for a client, but always to include within the problem-solving an aesthetic surplus that adds something to a place that can only be sensed. A simple example would be that by planting aspen (*Populus tremula*), one would not only remove attention from unwanted noise nuisance (the utility issue), but also add an aesthetic value by introducing the attractive sound of trembling leaves (the amenity value of the aesthetic surplus). But the aesthetics, the felt and sensed qualities, of the design that is communicated through the collage form only part of the complete picture. 'Collages cannot stand alone', Yoko once said to me, considering the difficulties associated with using them as communication tools with clients and stakeholders. Collages serve a great value as in-house work-in-progress tools enabling designers to focus passionately on the aesthetic elements of their design, but as external communication tools, their function is more questionable and uncertain.

Although intended primarily for in-house use, I would occasionally witness how SLA would nevertheless bring collages to meetings with clients, with unfortunate results. At one point, it came to a conflict where the client did not feel that SLA came up to the standard expected. SLA had presented collages in an 'identity analysis' which reflected a thorough observation of the history of a place, its distinct characteristics and the atmosphere to be incorporated into the subsequent design. What unexpectedly happened, however, was that the client was most annoyed by this and dismissed the work of SLA as 'fluffy'. The disappointed client confronted SLA in an e-mail:

> We always wanted you to go into the task and explore different options [in relation to ground plan, heights, parking spaces, etc.]. So why do you choose to prioritise vision, concept and identity over sketches? And this even directly after we've pointed out our need for *no* overall narrative. [...] It's very difficult for me to understand why you've failed to investigate what we ask of you [...]. All we need are good solutions. Nothing else. And we must be sure that these *are* good solutions and not just a random line.

The experienced team leader who faced the music of this outcry shook his head and explained to me:

> [The client] doesn't get it. This is how we always work, so why choose us for the task if [the client] isn't happy about our procedure? We always do identity analysis but perhaps we should just not show it to the client.

I think the team leader is right: the client does not seem to 'get it', and perhaps at times the collage serves the practitioners more than the clients. From a client's point of view collages can seem insignificant, abstract, random, impossible to 'read' and therefore 'fluffy', mainly because they do not address a call directly and in its entirety. As Yoko said, 'collages cannot stand alone', but need to be complemented by other ways of addressing space. Hence, in cases where the aesthetics of urban space are not an explicit issue for the client, designers will probably find a safer marketing strategy by addressing the 'situational space' before going into detail with all its aspects separately. 'In principle', the creative director once shared with me, 'the client is only interested in the utility. It is therefore the rational reasoning that we should employ when addressing our clients'. Not that the aesthetics – the amenity value of a project – should be concealed in the dialogue with clients, but it appears that a client's expectation is best addressed in its entirety. Collages, however, help designers to communicate about their overall vision for a project – its concept – which constitutes the crucial foundation of a project. Nevertheless, the collages are selective, sometimes even 'personal', they are 'edgy', and unlike renderings, they do not really care about others' wishes – neither the client's, the public's, or the designers'. Collages do not obey but, as we have seen, they ask questions. At some point I talk to Eric, an anthropologist, about SLA's dialogue with 'locals', and he explains that most architects in the studio think it's a bit tiring and 'folksy' to have people of various opinions too involved.

You need some form of friction to create something new, which touches on architects' self-understanding as being almost artists. There is a cliquish kind of aesthetic among architects, where they practically communicate more with each other than with others. Obviously you want recognition from your colleagues and not from the municipality or the local people.

Friction, Eric seems to suggest, does not come from consensual dialogue with clients and citizens of different convictions, beliefs, or sentiments. Friction, resistance, questions and 'edge' comes instead from the collage. And the collage clearly serves as an epistemological contribution in this respect. Additionally, when collages that have mainly been used for the conceptual in-house development of projects land on SLA's homepage at the very end, they might also be said to hold a social signal value of promotional use – in order to impress external colleagues too with SLA's artistic approach to urban design.

Conclusion

This chapter demonstrates that there is knowledge to be gained from a differentiated understanding of CGI. The mediation of urban sites by different kinds of images varies, and at a time when there is much controversy about the way designers visualise their ideas for the future, it becomes important to better distinguish collages from renderings and other types of drawing. I have here specifically focussed on the difference between collages and renderings – both of which aim to evoke atmospheres, but they address different audiences with different types of atmospheres that are more or less easy to understand for the uninitiated.

Renderings are produced with the intention of being easy to digest and they must hold a broad appeal. Aimed to humour what SLA employees would often call a 'many-headed client', insinuating the monstrous phenomenon of being confronted with the wishes of several clients at once, professional renderings would need to assuage potential worries or concerns of all parties, be they citizens, politicians, traders or proprietors, by providing answers, and thus ensuring appeal to the greatest common denominator. The role of the collage in urban design practice is very different from a rendering. From the outside and at first glance, the collage may seem insignificant. Collaging atmosphere is therefore easily disqualified as an artistic 'fluffy' excess that does not answer a call directly. A practical implication of this would be the cautious handling of collages in marketing, which, as the paper has shown, can turn into a risky business.

The collage serves several purposes, but first of all it is an epistemological tool that deals with uncertainties in the design process that are crucial to the design development. It serves as a 'rough' sparring partner that does not please but challenges the designers in order to support the innovate design. Collaging atmospheres also relies on skilled physiognomic landscape communication, based on a designer's ability to carefully select – take apart and combine anew within the collage – the key materials of a space that can make even an outsider feels its atmospheric vibe. Designers' physiognomic exploration of space requires their skilful attendance to its multisensorial appearance. Hence, collaging atmosphere makes vision work together with other senses in order to produce 'the stock cube' of a place – a metaphor that serves as a reminder about how the 'quick and easy' selective and fragmented way of making a collage can serve as a 'flavour enhancer' that intensifies and develops the designer's sense of a place in terms of atmosphere.

The implications of the discussion raised by this chapter are both theoretical and practical. To begin with the former, at a time when there is much controversy about the way architects and designers visualise their proposed ideas, it seems necessary to distinguish between the wide range of visual communication tools employed by designers, and here I have highlighted the role of the collage.

What is striking about how designers collage atmospheres is their skilled way of seeing – seeing how constituent parts of landscapes can touch, literally and figuratively, in order to suggest an overall feel of a place. From the outside and at first glance, the contribution of a collage may seem insignificant because is addresses only selective features of urban space. Collaging atmosphere is therefore easily disqualified as an artistic 'fluffy' excess that does not answer a call directly. As I have demonstrated, however, collages serve the design development at crucial points, and most significantly they add value by allowing the designers to engage in questions that can be asked and studied and not dismissed. Within this 'free' and explorative space of the collage, the designers can afford to focus explicitly on atmosphere.

References

Anderson B and Ash J (2015) Atmospheric methods. In: Vannini, P (ed) *Non-Representational Methodologies: Re-envisioning Research*. Abingdon: Routledge, pp. 37–50.

Barthes R (1981) *Camera Lucida: Reflections on Photography*. London: Macmillan.

Biehl-Missal B (2013) The atmosphere of the image: An aesthetic concept for visual analysis. *Consumption, Markets and Culture* 16(4): 356–367.

Böhme G (1995) *Atmosphäre. Essays zur neuen Ästhetik*. Frankfurt am Main: Suhrkamp.

Böhme G (1999) Die Physiognomie einer Landschaft. *Geographische Zeitschrift* 47: 98–104.

Böhme G (2001) *Aisthetik. Vorlesungen über Ästhetik als allgemeine Wahrnehmungslehre*. Munich: Wilhelm Fink.

Böhme G (2006) *Architektur und Atmosphäre*. Munich: Wilhelm Fink.

Böhme G (2017) *Atmospheric Architectures: The Aesthetics of Felt Spaces*. London: Bloomsbury.

Burrows R and Beer D (2013) Rethinking space: Urban informatics and the sociological imagination. In: Orton-Johnson K (ed) *Digital Sociology: Critical Perspectives*. London: Palgrave Macmillan, pp. 61–78.

Degen M, Melhuish C and Rose G (2017) Producing place atmospheres digitally: Architecture, digital visualisation practices and the experience economy. *Journal of Consumer Culture* 17(1): 3–24.

Dodge M, Kitchin R and Zook M (2009) How does software make space? Exploring some geographical dimensions of pervasive computing and software studies. *Environment and Planning A: Economy and Space* 41(6): 1283–1293.

Gibson JJ (1979) *The Ecological Approach to Visual Perception*. Boston: Houghton Mifflin.

Grasseni C (2004) Skilled vision: An apprenticeship in breeding aesthetics. *Social Anthropology* 12(1): 41–55.

Hasse J (2012) *Atmosphären der Stadt. Aufgespürte Räume*. Berlin: Jovis Verlag.

Hasse J (2014) *Was Räume mit uns machen–und wir mit ihnen. Kritische Phänomenologie des Raumes*. Freiburg: Karl Alber.

Houdart S and Minato C (2009) *Kuma Kengo: An Unconventional Monograph*. Paris: Éditions Donner Lieu.

Humboldt AV (1806)Ideen für einer Physiognomik der Gewachse. Tübingen. Available at: www.deutschestextarchiv.de/humboldt_physiognomik_1806/11 (accessed November 2021).

Humboldt AV (1844) Kosmos. Entwurf einer physischen Weltbeschreibung. Bd. 2, Hanno: Darmstadt.

Jackson M and della Dora V (2009) 'Dreams so big only the sea can hold them': Man-made islands as anxious spaces, cultural icons, and travelling visions. *Environment and Planning A* 41(9): 2086–2104.

Kaika M (2011) Autistic architecture: The fall of the icon and the rise of the serial object of architecture. *Environment and Planning D: Society and Space* 29(6): 968–992.

Kinsley S (2014) The matter of 'virtual' geographies. *Progress in Human Geography* 38(3): 364–384.

Kitchin R and Dodge M (2011) *Code/Space: Software and Everyday Life*. Cambridge: MIT Press.

Lund NO (1990) *Collage Architecture*. Berlin: Ernst & Sohn.

Graham, S and Marvin S (2001) *Splintering Urbanism: Networked Infrastructures, Technological Mobilities and the Urban Condition*. London: Routledge.

Melhuish C, Degen M and Rose G (2016) 'The real modernity that is here': Understanding the role of digital visualisations in the production of a new urban imaginary at Msheireb Downtown, Doha. *City & Society* 28(2): 222–245.

Morselli E (2019) Eyes that hear. The synesthetic representation of soundspace through architectural photography. *Ambiances. Environnement Sensible, Architecture et Espace Urbain* (5), doi: https://doi.org/10.4000/ambiances.2835

Pallasmaa J (1996) *The Eyes of the Skin: Architecture and the Senses*. London: Academy Editions.

Pallasmaa J (2000) Hapticity and time: Notes on fragile architecture. *The Architectural Review* 207(1239): 78–84.

Pallasmaa J (2005) *The Eyes of the Skin: Architecture and the Senses*. London: Academy Editions.

Pred A (1995) *Recognising European Modernities: A Montage of the Present*. London: Routledge.

Rose G, Degen M and Melhuish C (2014) Networks, interfaces and computer-generated images: Learning from digital visualisations of urban redevelopment projects. *Environment and Planning D: Society and Space* 32(3): 386–403.

Rose G, Degen M and Melhuish C (2016) Dimming the scintillating glow of unwork: looking at digital visualisations of urban redevelopment projects. In: Jordan S and Lindner C (eds) *Cities Interrupted: Visual Culture and Urban Space*. London: Bloomsbury.

Schmitt S (2018) Making charismatic ecologies: Aquarium atmospheres. In: Schroer SA and Schmitt SB (eds) *Exploring Atmospheres Ethnographically*. London: Routledge.

Schmitz H (1994) *Neue Grundlagen der Erkenntnistheorie*. Bonn: Bouvier.

Shields JA (2014) *Collage and Architecture*. New York: Routledge.

SLA (2021) https://www.sla.dk/studio/ (accessed 21 December 2021).

Stenslund A (2017) The harsh smell of scentless art: On the synaesthetic gesture of hospital atmosphere. In: Schroer SS and Schmitt SB (eds) *Exploring Atmospheres Ethnographically*. London: Routledge, pp. 153–171.

5 RENDERING ATMOSPHERES
THROUGH LIGHTING AESTHETICS

This chapter analyses how the perceived atmosphere of a place is communicated and rendered among urban designers, external collaborators and graphic visualisers. Urban designers may have an idea about how a place should *feel*, but how is this feeling communicated in renderings to make others feel the same or at least understand the vision for the atmosphere? The chapter illustrates how something as visceral as atmosphere is communicated and negotiated through the initial sketch proposal phases of a construction project. By following the process of making renderings, the chapter observes how designers communicate phenomena such as material elements, light and shadow in order to visualise atmosphere. The work is characterised by skilled ways of communicating about the intuitive flair of cultural aesthetics – in this case, *wabi sabi* beauty – where commonly used yet vague words are understood in rather precise ways among in-house designers yet travel with less ease across work cultures.

Feedback rounds for renderings for The Tunnel Factory

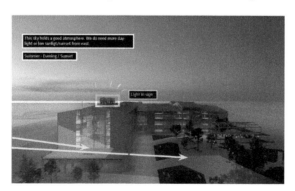

Image 5.1 Commented rendering View B early in the process. Bird's-eye view of the casting hangar and outdoor front area. A white 3D building model with a few lights and trees added. Comments point out what SLA and Arcgency want changed or approve of.

Illustration: Doug and Wolf, SLA and Arcgency

DOI: 10.4324/9781003279846-5

From: Matthew <matthew@dougandwolf.com>
Sent: Juli 3 2019 10:56
To: Evelyn <evelyn@architect.sla.dk>
Cc: Stenslund <ast@ sla.dk>
Topic: Re: Tunnel Factory | 3D model + Render | Handover

Hi Evelyn,

Please find attached few updated previews.

Concerning the aerial view (View B, Image 5.1). This one is a bit tricky.
We started with a dusk view which now progressively switch to a sunset view with
your last comments.
This can be a problem when at the end everything is lighten: inside and outside.
It usually leads to quite 'dull' or 'flat' result. That's why we did two tests here.
One is a real dusk when light can only be artificial. It's the better solution to
emphasize the inner light coming from the building.
The other one is a sunset. The last sun rays hit the ground and cast long shadows.
That's the best we can do to keep the building lighten from inside. Earlier than
that will become day view. And the building glazing will have to be darker than
the façade.

I hope it makes sense. Please let us know which one we should go with. (In our
opinion the sunset one is nicer).

Thanks.

The email is taken from the commentary rounds between Evelyn, a lead architect
in SLA, Denmark and Matthew, a graphic specialist in an Australian rendering
company. It marks a central point in the making of renderings for a new cultural
hub in a stretch of wasteland within the North Harbour district in Copenhagen.
After receiving the email from Matthew, Evelyn turns to her collaborators:

From: Evelyn <evelyn@architect.sla.dk>
Sent: Juli 3 2019 11:08
To: Lucy <lucy@collaboratingarchitect.dk>, Pete <pete@collaboratingarchitect.
dk>, George <george@designagency.dk>
Cc: Stenslund <ast@.sla.dk>
Topic: FW: Re: Tunnel Factory | 3D model + Render | Handover

Hi all,

Please find update on VIEW B. I expect other views ticking in tomorrow morning.
@Lucy Please see Matthew's comments regarding the birdeye view. We seem to
want a bit too many things with the image, which results in a flat image regarding
the light. Feel free to give me a call when you've taken a look at it.

Please submit your comments on the images in InDesign before noon.

/Evelyn

Comments from Lucy and other collaborators are then sent via WeTransfer, where it is noted that, 'View B is good, but we agree with you that there should be a different light. The picture here looks like morning light. It should be afternoon so there is light on the casting hangar'. Based on this input from Lucy, Evelyn returns to Matthew with her recognition of selecting the bird-eye View_B_Sunset due to the improved lighting conditions.

With this peek into the myriad exchanges of ideas and values in the process of creating renderings, this chapter explores the socio-cultural dynamics and aesthetic intentions in urban design that emerge between architects and graphic designers. The example above illustrates how questions are continuously raised about *what* and *how* components should be present in the renderings 'to depict and present specific embodied regimes and affective sensory experiences to appeal to clients and consumers' (Degen et al., 2017: 7). Renderings are productions aimed to pitch a masterplan to an external audience in the early stages of a design process, and they can serve as presentation tools throughout the construction process and for the submission of tender (Melhuish et al., 2016: 228). But as renderings aimed to evoke bodily feelings – that is, atmospheres, the question arises: what role do atmospheres play in the *process* of rendering prospective urban spaces?

As argued in Chapter 4, CGI, of which renderings are but one kind, may evoke 'digital atmospheres' (Degen et al., 2017: 4). Hence, attending to the way these images circulate as key interfaces in global, transnational networks of people and places, Melhuish et al. (2016) considers how they serve as 'vital platforms for communication and negotiation among producers and audiences' about cultural heritage and distinct urban identities, thereby paving the way to a new, digitally enabled, re-negotiated and postcolonial urban design aesthetic. Along the same lines, Degen et al. (2017) explore the aesthetic power relations that underpin CGI as cultural products, which in their study implies a premise of Western sensibility, even in their attempts to be culturally specific. While in line with these views that rest on perspectives from an Actor Network Theory approach (e.g. Degen et al., 2017; Melhuish et al., 2016; Rose et al., 2014; Yaneva, 2009), the analysis here is anchored in ethnographical work and guided by a phenomenological understanding of atmosphere as *felt* and *sensed* and far from always visible to the eye – neither in images nor in designers' actions or 'practices'. Following Gernot Böhme's (2017) emphasis on 'atmospheric architecture' including landscape design, I therefore challenge the visual representation of architecture and architectural image making in favour of its felt and sensed quality as atmosphere. My focus on atmosphere in rendering processes enables a perspective that, following Ingold, locates 'the creativity of the design process not in the exceptional faculties of "creatives", such as architects and designers, but in the generative potential of the social relationships in which all participants are involved' (2022: xvi). This chapter will show how atmosphere is not only the 'effect' of renderings *on* an audience, as Degen et al. have shown (2017), but acts as the kit that ties the whole rendering practice together.

Sunrise, sunset, dawn, daylight, twilight, dusk – decisions on proper ambient lighting in renderings, as illustrated by the thread of emails above, are of course not *only* a matter of taste or arbitrary opinions about what might flatten or increase the appeal of a rendering. The tangled affair of power and profit is not to be excluded from aesthetic rendering. For instance, in the rendering of View B (Images 5.1 and 5.2), a minor battle was played out between two groups. On one side there was Arcgency, the team of architects who designed the building complex (the casting hangar) and who would like to have it illuminated from the inside (best during the late afternoon or evening) in order for it to become eye-catching. On the other side, the multifarious group of urban designers in SLA who were responsible for the outdoor area and thus the building's surrounding environment – the 'context', they would say – were obviously interested in seeing their contribution as well. Afternoon or evening would leave the landscape in the dark so they would prefer to have it appear in daylight. This case of conflicting interests between contributing collaborators tips into the theoretical discussion of building architects' tendency to 'decontextualise' urban design (Grubbauer, 2014), and it opens a door for studies of potential alliances

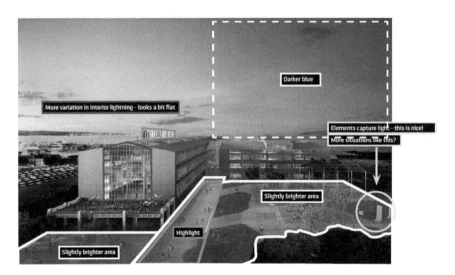

Image 5.2 Commented bird's-eye View B addressing the lighting at a later stage in the rendering process.

Illustration: Doug and Wolf, SLA and Arcgency

between architects, designers and financiers (Dovey, 2010). Seen in this light, literally and metaphorically speaking, architectural image-making becomes a crucial instrument in helping to visualise urban policy, and financial goals become easier to implement if they are visually represented in certain ways (Røe, 2015: 56).

The track I would like to pursue in this chapter does not lead into theoretical conceptualisation of power struggle and distinction, but introduces a less well-versed subject matter in design's sociological, anthropological and social science analysis in the widest sense: rendering as a way of engaging emotionally in urban design 'in, about and through' atmosphere (Sumartojo and Pink, 2019). Atmospheres are transformative by nature – they mark and penetrate all who take part in the making of renderings including those who seek to learn about it, right from the collaborating architects and designers to me in the role as ethnographer. This ability forces me to flag the need to move beyond clearly distinguished concept formation and to understand the design practice and my own conceptualisation of 'interwoven' in Ingold's sense of the term (2013). Just like the designer's ideas for the urban environment are not adequately understood to achieve their form by the moulding of a material supply (Ingold, 2013: 20), their practices, likewise, do not serve the social scientific study as an empirical supply to be moulded by predefined theory. This is the motivation for my enquiry on atmosphere in renderings. The questioning of atmosphere and aesthetics in design still receives less attention than more salient subjects of social power and distinction, but as this book argues, atmospheres are important and often underrated co-players in processes of developing urban design. This chapter illustrates how the aesthetics of lighting guides the rendering of atmosphere, and how urban designers do not merely create atmosphere through the application of personal taste or intellectually justified values but are also part of the atmosphere that grows from the rendering process itself. This takes the reader on a journey from how the perspective and focus in the rendering is negotiated through lighting, to the cultural appeal of light. Yet, before turning to empirical examples of making renderings for The Tunnel Factory and its surrounding wasteland, I need to address the very nature of renderings and their atmospheric impact.

What is (a) rendering?

According to the Cambridge English Dictionary (2021), a rendering (also: rendition) can be, for instance, a performance in the sense of 'the way that something is performed, written, drawn, etc.' It also means 'a translation of […] [a] piece of writing into a different language or a different style' and additionally it denotes 'a work of art or performance that represents something'. Similar definitions are given in the Collins English Dictionary (2021), which adds to these a definition in architecture as 'a perspective drawing showing an architect's idea of a finished building, interior, etc'.

The term 'rendering' thus has a double meaning as both the action of preparing a rendering and the result of the action. The performance of this action, 'to render', is worth exploring 'anthropographically'. 'Anthropography' in opposition to 'ethnography' is a term coined by Ingold (2013: 129) in order to describe a fieldworker's sense for transformation processes (contained in the verb to 'render') instead of 'pure' descriptive observation.

In architecture, renderings are detailed visualisations of design solutions for prospective projects, and colloquial language during fieldwork in the studio suggests that renderings are often simply called 'visualisations': 'I don't say "rendering"', an architect corrected me and carried on, 'of course you can say "rendering", but "rendering" is just a reminiscent term from V-ray – a Rhino plug-in [for a visualisation program] – and thus rendering is just a way to talk about something when it has received an extra treatment'. Hence, when I asked about rendering, many, such as this particular architect, would understand rendering as *a* visualisation, which is mirrored in some of the quotes that I include below. Yet throughout my conversations with employees in the studio, it also becomes clear that 'rendering' appears both as a noun and verb signifying a process when something is *done* – or given an extra treatment, as noted. It is also present in the subject of the email thread illustrated above: '3D model + Render | Handover'. For an outsider, 'visualisations' can mean almost any kind of image creation. I use the notion of rendering, then, not as a clear emic expression (not to be confirmed from an insider's perspective), but an etic, analytical one (where I consider the studio culture as a researcher from the outside). This allows me to address how multifarious atmospheres permeate a rendering (noun) and processes of rendering (verb): a rendering serves to give the viewer a sense of the desired atmosphere of a redevelopment project's end result, which again can be the result of atmospheres enveloping the rendering process itself – processes that are normally hidden and inaccessible for those who are not part of the design studio.

Renderings serve a communicative role throughout the entire design process, but first of all they are created to win project contracts. They are produced in the initial phase of the design process to convince the client that the absolute best project proposal is revealed in and by the rendering. Renderings thus need to be 'thoroughly well-worked-out images that tell the client that this is how [a project] will look, and [the rendering] must look like reality', another informant said, confirming that the room for translation is narrow and not the perceivers' obligation. While reflecting reality as well as possible, a rendering also keeps the dialogue about a project's specific solution on grounds of principle so as to avoid breeding too many specific expectations at an early stage, such as geometry, proportions or choice of materials. This balancing act between providing realistic impressions of a not yet realised project is supported by the focus on the atmospheric effect of the rendering. Renderings must seem realistic, architects would tell me, but not 100% accurate in every detail. In the studio, good renderings translate the design vision into an image in a credible way, allowing clients, developers and user groups to feel and recall the overall atmosphere that a project seeks to create. In other words, with the rendering, urban designers strive to make any outsider able to envision the end result. Renderings therefore do not communicate about an objective state of affairs but phenomenological lifeworld notions about what it might feel like to be in a particular place in the city. One architect's comments on a specific rendering are telling when he states:

> The things that are green here are also green in reality – but this [issue about colouring] is really up to the individual aesthetics of a company. In principle, you could also make the illustration pink – pink tree trunks and so on. It can work just as well and is not really an issue – it has more to do with personal style and taste.

As the introductory feedback round also illustrates, renderings perform atmospherically rather than being factual as such. This means that what matters is how their felt presence reveals itself at an experiential level rather than how they look objectively. They should not show 'flat' or 'dull' results but the opposite: they are to 'seduce as many people as possible', a lead architect told me once. Gernot Böhme points out the difference between *factual fact* and *actual fact* (*Realität* and *Wirklichkeit* in German) (2001: 57). For instance, it may be that factually a room is a certain size, but *actually* it may feel much smaller or bigger than it really is according to its measurements. Likewise, in *fact* there may be certain colours in the rendering but they may well be perceived differently due to the *actual fact* of their atmospheric quality. So, whether the tree trunks are in fact brown, pink or purple is truly a 'real matter of fact', but it is still inferior to their atmospheric quality of the rendering, which is a question about their phenomenological appearance to the viewer.

A transformation in architecture firms from mainly providing functional solutions to becoming more experience-oriented (Klingman, 2007) chimes with this phenomenological approach of urban designers. In the studio it was not uncommon to come across phone calls with clients broadly paraphrased as: 'Yes, we understand that you would like this building removed, that building restored, parking space suited for x number of cars and also you ask for a solution to the noise nuisance over here. We can easily find a solution to all of it, but can we please ask: how should it be for people to stay there? How should the area be experienced?' This felt, sensed and lived quality of a place is explicitly cherished by the studio. In SLA they would set off with an idea about how a place should feel in terms of its vibe, and this nitty-gritty heart of the matter would then be settled as the 'concept' or principal idea of a design solution. As my colleague Mikkel Bille, with whom I wrote a first edition of this chapter, reminded me, this is reminiscent of Adolf Loos' call for the architect to first 'identify a feeling for the effects he wants to create' (Pérez-Gómez, 2016: 20). Within the studio 'the felt and sensed' (in Danish, *det mærkbare*) refers to 'the atmospheric' and rendering (verb) is about 'grasping' and 'capturing' this atmospheric touch of a project and passing it on to a larger audience via the rendering (noun). To illustrate I now turn to the conceptualisation of a redevelopment plan for The Tunnel Factory, a huge wasteland area in Copenhagen's North Harbour.

The Tunnel Factory

A massive industrial building standing in the outer harbour area in Copenhagen is to be transformed – according to the client's wish – into a brand new sustainable hub where creativity, inclusive communities and lifestyle urbanism are at the forefront. Today the building is renamed The Tunnel Factory in order to embark on a new identity. It will no longer serve as the former ÖTC factory, built in the 1990s to mould the tunnel elements for the Øresund Bridge linking Sweden and Denmark. The developers' new vision is to transform the post-industrial building and surrounding outdoor area into a complex of arts and culture scenes, workplaces, places to eat, studios, workshops, sports facilities, leisure, entertainment, city nature experiences and much more. In order to bring these visions to life, renderings are needed, and SLA takes the lead in this task. They have hired Doug and Wolf, an Australian company specialising in renderings. This choice of collaborator is due to the studio's impression of the delicate sensitivity that Doug & Wolf has so far shown towards the aesthetic preferences of their clients: 'When our designers ask for "wild nature", Doug and Wolf doesn't suddenly plant placenta flowers in the renderings', Evelyn explains to me, indicating that this aesthetic slip was made by other less successful collaborators.

Aesthetics here enables me to approach the strong economic agenda that influences the process of urban design like any aspect of our environment (Friberg, 2019). It is important for the real estate provider to immediately attract the creative class (Florida, 2005) in order to have them start renting spaces to help recoup some of the construction costs while simultaneously contributing to the liveability of the area. Hence, it is SLA's job together with its collaborators to have the renderings address the exact values and 'good taste' of the target group. Rendering, thus, is clearly about making an area appeal and make it an object of social distinction (Bourdieu, 1984). For instance, at meetings in the studio I would witness discussions on how to create an image of this hub that in reality was available to everyone, but people would still feel among the few who knew. The image of a secret spot would be upheld not only through the rendering, but also through the choice of (no) advertisements, the sale of natural (unadulterated) wines, and by the arrangement of 'public' readings of literature – activities and programmes that the urban designers would imagine appeal to the social group that would generate money and innovation as a commodity (Røe, 2015).

The location of Doug and Wolf in Australia was a timely advantage to SLA, who could exploit the working hours' time difference. Guided and commented rendering-editions would never be at a stand-still but be processed 'down under' while collaborators in the North were asleep, and vice versa. Also, it seemed to be advantageous to the production of renderings to choose a partner who is miles away from the social struggles that might take place between stakeholders and design collaborators within the Copenhagen area. Detached from local conflicts of interest, the external rendering company is hired partly to just take note of interests explicitly formulated in a pre-set 'rendering guide' formulated by SLA and collaborators, but partly also in order to specialise in 'good', appealing, credible rendering techniques with aesthetics at the forefront. It comes as no surprise that companies will have an interest in promoting awareness of their own originality (in this case, to have the construction work or the surrounding landscape stand out in the light), yet it is remarkable how the two Danish companies, in this case SLA and Arcgency (the architectural designers), consent wholeheartedly to external advice from Doug and Wolf that deals with the atmosphere of the rendering and that mainly addresses its aesthetic appeal and not the self-interests of the companies.

The atmospheric hour of a rendering makes the eye rest

Atmospheric renderings are composed to appear simple. Good renderings, Evelyn tells me, usually display just one situation and refrain from asking the recipient to relate to several things at once. It sounds straightforward when she tells me this, but when following the process, it seems to be less easy, especially in cooperative situations. Rebecca, an architect in the studio notes that:

> Sometimes we have to be better at showing the simple picture. We often want to show everything at once; space for children, the elderly, bicycle parking, etc. But all of a sudden, the atmosphere becomes difficult to see, even if there is a beautiful light, because [the recipient has] to take a stand on way too many things. [...] It's something we're quite aware of that we need to improve. We don't have to show everything, but this is nevertheless what we sometimes feel an urge to do, since of course you want to make sure that everyone in the judging committee or whoever receives it can see themselves [and their wishes reflected] in the project, rather than solely being seduced by a mood.

Rebecca exemplifies the key issue discussed in the feedback rounds: how to create an atmospheric impression that is simple enough to seize the viewer? For example, rendering comments show how lighting plays a crucial role for how simple the rendering appears, and thus how narrow or broad the focus should be for the person who sees the rendering. At meetings with developers, it was discussed what time of day would best represent the site. Since the bird's-eye View B (Images 5.1 and 5.2) was the intended 'money shot' to hit the headlines with the overall project proposal to turn The Tunnel Factory into Denmark's largest cultural hub to date, much was at stake in the rendering process of this view. Evelyn tells me how prior to placing the order with Doug and Wolf, she did some 3D-testing of how the site would make its best appearance according to the choice of time of day – that is to say, the lighting conditions:

> I made one for 10 a.m. and one for 12 a.m. and I've tested it in the bird's eye view. We had fancied the idea of a midsummer-day-having-a-good-time-like picture [Evelyn is snapping her fingers]. You can see over here that 5 p.m. does not work well because of the huge shadow thrown across the arrival area – how people arrive was really what we wanted to show. That's why we thought it was much better [to render The Tunnel Factory] in the middle of the day – so that we would be able to see both the arrival and the surroundings. But then this architect trumped us by saying that he didn't want daylight because then you wouldn't be able to see into the building. In daylight, windows are dark, and with the light reflection you can't see what's going on indoors. He wanted to show that there were 'days of wine and roses' inside the building – you know, the general idea is that it must look lively 24/7 with hipsters all over the place and then a more commercial shopping street with restaurants and so on. That's why we ended up ordering a picture at dusk, in order for there to be light both inside the building and just enough light outdoors in order for you to see what's going on.

Based on what and how much the viewer should be able to see, the studio decides to order a picture at dusk for there to be just enough light both inside and outside the building, considering interests of both indoor and outdoor urban designers. However, as the e-mail correspondence also reveals, the final rendering ends up being at neither dawn nor dusk. It ends up as a sunset image (Image 5.3). This is due to the disclaimer of Doug and Wolf, who flag their reluctance to meet everyone's wishes. Instead of spelling this out, they put it down to the lighting issue: the time of the day rendered in the image, they remind the employees in the studio, helps determine how many stories (or agendas) the rendering can contain. The sunset image is an alternative to dusk and dawn provided by Doug and Wolf who seem responsive to some of the original wishes formulated by SLA on behalf of their client – By & Havn, the corporation for development of City and Port I/S owned by the Copenhagen Municipalities and the Danish Government – who would like to show a coastline in the background. 'The coast is still lit up in the final rendering', Matthew writes back in response. At the same time, Matthew

Image 5.3 Final bird-eye View B+sunset with 'interesting' long shadows thrown across the arrival area. Outdoor and indoor activities are visible, as is the coastline.

Illustration: Doug and Wolf, SLA and Arcgency

points out to the Danish designers that a sunset picture will appear warmer (in reddish tones) and with longer shadows cast. 'So, in the end, we chose to go with his suggestion', Evelyn explains, 'because obviously the shadows add a greater play of light and make the image appear more interesting than the view at dawn'.

The light creates focus in the sense that it spotlights the conceptual storyline that ensures the atmospheric quality of the rendering – making it compelling, nice and easy, rather than confusing and complex due to too many details. Rendering atmosphere thus involves a designer's careful attention to what light allows an image to tell. 'A rendering should be read in a split second', another SLA architect tells me, and she clarifies that the correct lighting ensures that the eye instantly finds a place to rest. Rendering atmosphere hence allows for a perceptual hierarchy with a clear centre of attention and a less important periphery, which in this case prevents dullness and 'flatness' in experience, keeping a focus and ensuring simplicity by appearance. The point is then that the rendering acts through atmosphere, which again is framed through the interplay between light and storyline (what to see and not to see). This raises the question of the role played in the design process by cultural aesthetics and communication skills.

Cultural aesthetics of light, subtlety and cheesiness

For View A (Image 5.4), one of the other six professional renderings for The Tunnel Factory, Evelyn and her team ask for some of the light rays to be removed. 'It's getting too cheesy, Evelyn explains. 'It's almost a divine light. It's like "too much". It has to be more subtle. Something that doesn't steal all the attention and still works well'. When I ask what it is that works well, one of Evelyn's colleagues elaborates with reference to a rendering from another project:

> The light is like a goddess coming down. It's like come on, silly! Now Jesus himself is coming down from heaven! The person who fancied the visualisation wanted so much godliness that it looked like Jehovah was coming down, and we [at the studio] were just, like, 'chillax'. We prefer it to be a little more subtle. [...]

'What does subtlety do? I asked. 'It offers rest. After all, it's difficult to focus on what you've done – that is, what you've actually drawn – if all is covered in blazing sun rays.

In the above situation, it was a client from England who fancied the 'cheesy' sun rays, I was told, and from the viewpoint of Evelyn's team it was a matter of taste, which was exactly why the process of rendering could

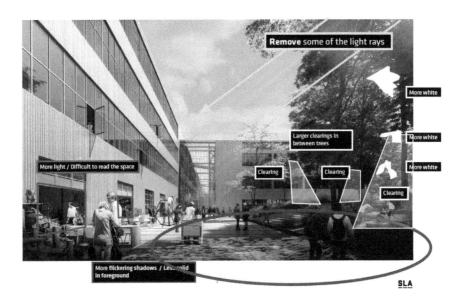

Image 5.4 Commented View A street level showing activities on the promenade with workshops in the building to the left and a green vegetated area to the right.

Illustration: Doug and Wolf, SLA and Arcgency

sometimes turn into a challenging task of manoeuvring among different cultures' styles, tastes and preferences.

This issue of the aesthetics of rendering becoming a cultural matter of taste is a general point that applies to most renderings. Although the studio does not conduct systematic segment analyses, it turns out that employees who work on renderings draw on an intuitive flair to ensure the atmospheric appeal of the renderings. For instance, Wei, a Chinese employee in the Copenhagen-based studio, qualifies this matter of taste by pointing to lighting adjustments in renderings. He distinguishes renderings for Scandinavians from renderings made for Southern European or Asian clients:

> I feel like in the Scandinavian countries, the weather is a key issue, because the sky is always kind of cloudy and grey. It's very rare to have like a sunny bright blue. [...] Actually, everything is kind of muted in the renderings for Scandinavia. I don't know why. It's just the kind of feeling you get when you walk in the street in Copenhagen. You feel like... it's not like colourful Spain [...] where there are really bright red and yellow colours. In the Scandinavian countries you don't use colour a lot [...]. It also goes with the personality of Scandinavian people. Everybody is shy, not very outgoing. People have their own inner peace; they are very quiet and don't have much expression on their faces or interactions with people on the street. Maybe this is why muted colours go with this kind of environment. When you're in a more colourful world, you feel more outgoing.

Wei is a good example of how architects and designers intuitively work with the impression that choice of colour, for instance, implies. Wei thinks of geographical regions in terms of colours loaded with sentiments that match identities of people and places – this resembles the physiognomic reading of space described in Chapter 4. Scandinavian people's sentiments captured in a muted colour spectrum confirms his aesthetic work with atmosphere producing renderings able to impinge on viewers with felt sensations of Scandinavia. By 'turning down' the colours, the designers would 'turn up' the atmosphere that they perceive exists in and around Scandinavia.

In order to render atmosphere, one would need to have a sense of both the space under redevelopment *and* the type of person to which the rendering should appeal. From the architects and urban designers, I hear of episodes where external rendering companies inserted improper elements into the images like a red bull, a severely obese person or a sinewy guy with tattoos. 'Such gimmickry just doesn't work', says Evelyn. 'It's not that we don't want to be diverse and inclusive, but the stand-in people that we use have to look like our target audience and they cannot capture the attention by breaking the norm'. As with the careful selection of suitable people for the renderings, the same careful selection is done regarding the activities that 'suitable' people are engaged in, suitable vegetation and choice of furniture, façades and paving. The cultural aesthetics of all of these mentioned features together constitute the overall 'material narrative' of a rendering.

The social mechanisms of rendering

In a feedback round for View E (Image 5.5), Evelyn asks the Australians to 'delete the flower stand'. 'It looks like a square from Albertslund Centeret [a suburban shopping area]. No offence, this is where I come from myself, but you know, this isn't where we're heading with this'. 'So, where are we heading?' I ask. 'In my opinion, this is going to be a polished hipster centre', says Evelyn, implying a particular lifestyle urbanism that their design choices need to support. So out goes the petty-bourgeois flower stand and in come long grasses – the ultimate hipster vegetation nowadays consistent also with *wabi sabi* aesthetics.

For this project, the architects and urban designers collaborated with ArtRebels – a 'cultural studio and creative community for unconventional thinking', as they describe themselves on Facebook. ArtRebels, who were not familiar with making renderings, were however expected to join the feedback rounds circling between the urban designers in the North and Doug and Wolf in Australia. Yet, Evelyn says, since they are untrained in the rendering communication, working with them is a challenge. That is

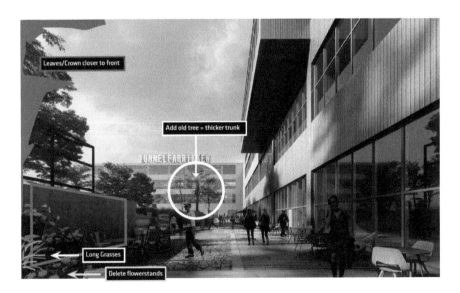

Image 5.5 View E. Street level from the opposite end of the promenade (compared to Image 5.4). Vegetation to the left and building to the right and straight out. Comments address the vegetation.

Illustration: Doug and Wolf, SLA and Arcgency

Image 5.6 View E with comments and reference pictures addressing materials. A café in the lower left corner is encircled and two reference pictures of alternative-looking bars are inserted to show the desired style that the designers would like to change the image to.

Illustration: Doug and Wolf, SLA and Arcgency

why Evelyn had to translate what this community meant with their forwarded reference images about which ArtRebels tended only to argue. Pointing towards View E in a version including comments on its materials (Image 5.6), she says:

> Well, I understand what they mean here [with recycled containers instead of a posh-looking bar], but if others should be able to understand, we still need to spell it out, literally speaking. We just can't assume that the Australians went to Reffen [a hip part of Copenhagen with faded shipyard grandeur] last Saturday, you know what I mean?

This challenge of engaging in a cross-cultural communicative action shows that in order to render, one would not only need to know about cultural norms, values, 'good taste' and *wabi sabi* beauty. One also needs to master communicative skills to formulate one's intuitive flair about space identity and atmosphere in written text. When Evelyn, in each feedback round, thus formulates a collective response dispatched from all her collaborators in Copenhagen to the Australian Doug and Wolf, she would literally translate images sent from ArtRebels into text. For example, she would spell out that 'the food stalls should not be made of concrete. We prefer remodelled shipping containers instead', and in yet another version of View E she would write that they wanted 'Berlin-type furniture, table + bench' instead of Hay design and Acapulco chairs, and as for the façade panels of the building, she would tell Doug and Wolf that they were to be 'reused' and not 'look too brand new' as was the case in the images. As Mădălina Diaconu has also noticed, patina has certain aesthetic connotations that are used as positive values 'aestheticisizing [the] effect of time' (2006: 131). Patina appears when surfaces have been repeatedly touched by human hand and climatic change and adds a 'temporal depth' attractive to the designers since it may be emotionally sensed as atmosphere.

So when urban designers based in Denmark ask for 'weathered', 'rubbed', 'used', 'reused', 'less polished' and 'less neat' materials in their feedback rounds for the rendering process, then it is not because they desire some rubbish bits and pieces, but because this choice of materials adds to the intended atmosphere and beauty of *wabi sabi*. Their choice of materials for the renderings are thus based on aesthetic 'unconventional' ideals and values which are in need of cross-cultural translation. The translation is not necessarily challenged by national boundaries, since Nordic *wabi sabi* aesthetics is presumably better understood by the Australian Doug and Wolf, who were better at 'breaking the code', than their European competitors located closer geographically. Rendering atmosphere is therefore less about cross-national communication than about skilled wabi sabi sensitivity and communication and translation of cultural aesthetics and taste.

Hue, saturation and vibrance

'Well done', Evelyn says to her team after agreeing on the final results of the renderings. Three team members have crowded around Evelyn's computer screen, all gazing at the images. The renderings for The Tunnel Factory will go out to clients and stakeholders in less than an hour 'and then we definitely deserve a beer', Evelyn perks up. 'They will just need a final touch', she says and grins. This grin might have passed by the uninformed bystander, but my long-term presence in the studio allowed me to witness how certain ways of completing the rendering processes would recur. After lengthy negotiations on every detail of the renderings, including colour, light and tone, in the studio they would nevertheless end by muting or desaturating final versions of renderings themselves.

Why they did not simply ask the external rendering companies to adjust the colours during the feedback rounds was a mystery to me. I asked several employees, including Stig L. Andersson, and his answer was that it probably had to do with atmosphere being difficult to communicate to people who are not part of it. But the atmosphere admittedly was not yet there, as such, it existed in the designers' imaginations, and they could just have asked for 'a little less colour' without any extra costs. There are different ways of explaining this, and many employees would tend to see it as a question of 'good' or 'less good' taste. Evelyn's colleague explains to me:

> When we receive the final visualisations from the external agency, there is no more that can be changed. It's [the external rendering company's] product as well – not only ours – and if we say that we want something different at this stage, we would then have to pay them in order to make further changes. However, often we upload the final visualisations into Illustrator [a vector graphics editor] and as a final touch we slightly adjust the colour. Just a tiny bit.

And why is that? I ask her. She really does not know, she says and smiles, 'it just looks nicer'. And then she shows me how the colour can be 'pulled out' of the image with three buttons in Illustrator: ''hue', 'saturation' and 'vibrance'.

A standard sociological and anthropological explanation would talk about appropriation, stressing the need for SLA's ownership turning a common product into their own. Such an explanation is consistent with an interview given by the Head of Doug and Wolf to the Danish press about the type of negotiations that take place during rendering processes: 'We are somehow artists working for artists. It's tricky. The architects have already gone through a complete creative process where they have had to deal with limitations [...], which is why they are not always super happy to hand over their designs to [our] team who are excited and ready to start their own creative process' (Jensen, 2019). Seen in this light, rendering atmospheres are not excluded from social power relations and negotiations of ownership.

Without excluding any of the possibilities given above, I would like to add a facet, because another possibility is that SLA mute the colours for their own sake and for the atmosphere and team spirit in the office regardless of the external rendering company's output. What I see is that they are indeed asking for the final adjustments (Image 5.7). Hence, they seem not to be reluctant to give the renderings a 'final touch'. But is it not conceivable that, no matter how precisely Doug and Wolf in this case hit the requested tone, it will never be good enough? Adjusting light and colour, saturation, hue or vibrance has found an almost ritualised format in the studios' rendering teams who huddle around a screen at the very end. What I argue is that it is not just the team's ability to work together internally and with external collaborators that enables them to render atmospheres in an image format: it is also the rendering as image creation that sets an atmosphere in the team, and this atmosphere can hardly be outsourced but exists in situations within the office only. The rendering process is thus about both an external communication and an internal point of reference for the further project, and also about individuals' social positioning in terms of gaining recognition internally.

Image 5.7 Final light adjustment of View B. Original image to the left with brighter test image to the right.

Illustration: Doug and Wolf, SLA and Arcgency

Conclusion

One major issue in social scientific dealings with architectural renderings has been to discuss the 'affective' appeal of clients and consumers. This chapter illustrates how lighting aesthetics guides the rendering of atmosphere, and how urban designers do not merely create atmosphere through the application of personal taste or intellectually justified values but are also part of the atmosphere that grows from the rendering process itself. It has argued that rendering atmosphere involves a unification of light, materiality, architects, graphics and designers that frame what the viewer must be able to perceive and thus feel on experiencing the rendering for the first time. As an image-making process, renderings involve an atmospheric competence of curating aesthetics that rests on intuitive flair and skills of communication. Rendering as noun and verb is interwoven with selling dreams for the future that are credible and not too unrealistic by turning the question from *what you see* to *how you feel* about prospective urban spaces. I have shown how the rendering process is not only involved in cross-cultural negotiations where atmospheres are attributed to the rendering in favour of clients' visions for future cityscapes. It is also about how atmospheric sensory appearances mark the very materialisation process of the renderings as well as the communication between participating collaborators. The atmosphere is, so to speak, the 'glue' that ties the creative process together. More than simply a noun, rendering is a verb where atmospheric competences are negotiated and moulded in a process that in particular entails choosing the light that makes future places not just look but also feel a particular way.

References

Böhme G (2001) *Aisthetik: Vorlesungen über Ästhetik als allgemeine Wahrnehmungslehre.* Munich: Wilhelm Fink.

Böhme G (2017) *Atmospheric Architectures. The Aesthetics of Felt Spaces.* London: Bloomsbury.

Bourdieu P (1984) *Distinction: A Social Critique of the Judgement of Taste*, translated by R Nice. Cambridge: Harvard University Press.

Cambridge English Dictionary (2021). https://dictionary.cambridge.org/dictionary/english/rendering (accessed 23 December 2021).

Collins English Dictionary (2021). https://www.collinsdictionary.com/dictionary/english/rendering (accessed 23 December 2021).

Degen M, Melhuish C and Rose G (2017) Producing place atmospheres digitally: Architecture, digital visualisation practices and the experience economy. *Journal of Consumer Culture* 17(1): 3–24.

Diaconu M (2006) Patina – atmosphere – aroma: Towards an aesthetics of fine differences. *Analecta Husserliana* 92: 131–148.

Dovey K (2010) *Becoming Places: Urbanism, Architecture, Identity, Power.* London: Routledge.

Florida R (2005) *Cities and the Creative Class.* New York: Routledge.

Friberg C (2019). To answer a demand: Aesthetics in economy. *Studi di estetica* 15, doi:10.7413/18258646097

Grubbauer M (2014) Architecture, economic imaginaries and urban politics: The office tower as socially classifying device. *International Journal of Urban and Regional Research* 38(1): 336–359.

Ingold T (2013) *Making: Anthropology, Archaeology, Art and Architecture.* London: Routledge.

Ingold T (2022) Foreword. In: Stender M, Bech-Danielsen C and Hagen AL (eds) *Architectural Anthropology: Exploring Lived Space.* London: Routledge.

Jensen RH (2019) Doug and Wolf sætter spørgsmålstegn ved brugen af romantiske viualiseringer. *Byggeri + Arkitektur.* 2019.06.12. Online magazine: https://byggeri-arkitektur.dk/doug-and-wolf-saetter-spoergsmaalstegn-ved-brugen-af-romantiske-visualiseringer

Klingman A (2007) *Brandscapes: Architecture in the Experience Economy.* Cambridge: MIT Press.

Melhuish C, Degen M and Rose G (2016) 'The real modernity that is here': Understanding the role of digital visualisations in the production of a new urban imaginary at Msheireb Downtown, Doha. *City & Society* 28(2): 222–245.

Pérez-Gómez A (2016) *Attunement: Architectural Meaning after the Crisis of Modern Science.* Cambridge: MIT Press.

Røe PG (2015) Iscenesettelser av den kompakte byen - som visuell representasjon, arkitektur og salgsobjekt. In: Hanssen GS, Hofstad H and Saglie IL (eds) *Kompakt byutvikling. Muligheter og utfordringer.* Oslo: Universitetsforlaget, pp. 48–57.

Rose G, Degen M and Melhuish C (2014) Networks, interfaces and computer-generated images: Learning from digital visualisations of urban redevelopment projects. *Environment and Planning D: Society & Space* 32(3): 386–403.

Stender M (2017) Towards an architectural anthropology: What architects can learn from anthropology and vice versa. *Architectural Theory Review* 21(1): 27–43.

Sumartojo S and Pink S (2019) *Atmospheres and the Experiential World: Theory and Methods*. London: Routledge.

Yaneva A (2009) *The Making of a Building: A Pragmatist Approach to Architecture*. Oxford: Peter Lang.

6 SCALING ATMOSPHERES

INTUITIVE DIGITAL ZOOMING IN AND OUT ON MATERIALITY

Scaling – the technique of working with the proportional size and organisation of space – is one of the fundamental tools for designers when working on landscapes, buildings and products. It would be easy to think of scale as a purely quantitative discipline involved in the algorithmic processing of space geometry, but ethnographical research among architects has already shown how scaling is in fact also practised with rhythm of movements that edit, adjust and size up space (Yaneva, 2005). Observations presented in this chapter add to the existing ethnographical literature by rethinking scale even further in terms of atmosphere. Digital scaling, I propose, involves empathy in potential emotionally charged experiences of future space that are picked up in the designers' imagination and integrated into the design – not only to captivate users of future urban spaces but also to captivate and make the designers feel connected. In addition to contributing to how atmosphere plays a role in-house among designers, there is also knowledge to be gained about how qualitative and quantitative elements of design are infiltrated rather than being mutually exclusive. Hence, something happens in qualitative terms all the while future urban space is approached through measurement.

Moving digital scaling

'So first of all, I have to get an idea of the proportions', says Selma. She inserts a human silhouette – a scale model like a clipart graphic – into her Rhino model. 'I kind of need to get an idea of what we're dealing with', she elaborates. The way Selma forms that idea is by measuring the area and then she locates the 'fixities' (existing buildings and trees that cannot be touched) in order to assess how much space they take up relative to each other. Without scale models, like the human silhouette, space is too abstract and difficult to grasp. People create a bodily feel for their surroundings and it seems that designers are no exception to this rule. From Yaneva we learn how architectural scaling is in fact a very tangible and physical affair where designers 'jump' back and forth between upscaled and downscaled mock-ups by entering a dialogue with their materials and shapes (2005). However, scale modelling, as studied

DOI: 10.4324/9781003279846-6

by Yaneva, is not the same as the digital kind of scaling to be discussed in this chapter. They are interrelated disciplines that 'use different strategies, techniques and methods to achieve the same goal – the original and quality presentation of an architectural and urbanistic work to a prospective client/audience' (Stavric et al., 2013: 19). Models in terms of mock-ups hold a clear, tangible physical form that allows people to walk or jump around and engage with them from various angles, whereas computer modelling and scaling happen virtually with a few clicks – and a lot of scrolling, as we will see. What I suggest, therefore, is that scaling not only involves designers who 'move around' in order to learn about constructions or landscapes under redevelopment; scaling also moves the designers in terms of mood and makes them connect with particular projects.

The atmosphere sheet

When Selma and her team are about to figure out how a space should be organised and what features it should contain, they attend to project programming. Architects and designers speak about 'programmes' when they are to identify the scope of work discussed and agreed during the initial design stage. The programme identifies the client's needs and wants – for example, it could be problems with the terrain, which needs to be easier to navigate or there may be a desire to facilitate certain activities, etc. But often, in parallel, Selma does what she calls an 'atmosphere sheet' or sometimes 'atmosphere palette' – a piece of paper or a blank document that pins down a suggested atmosphere that she and her colleagues will aim for throughout the rest of the design phase (Image 6.1). Selma stresses to me that doing these atmosphere sheets is her own preferred way of starting a project, and she does not know what 'everybody else is doing'. It is therefore not a common, generally implemented method I am about to describe, but a tool that *can* be used by designers. In SLA designers are free to find their own preferred working methods. Other methods I identify in use that involve atmosphere are collaging (described in Chapter 4), making mood boards or the 'identity analysis' (*egenartsanalyser*). The reason I find the atmosphere sheet particularly interesting is, first, its explicit interest in atmosphere and, second, it is a method based on scaling. And since working with scale is usually considered a quantitative method only, not associated with atmosphere, it calls for more thorough investigation.

In order to do the sheet, Selma needs to familiarise herself with the space under redevelopment. Familiarising means here to 'get into' the place. Her literal wording is '*jeg sætter mig ind i stedet*', which means to familiarise oneself with or enter a space. In direct translation she says 'I sit myself in the place' by paying an 'ultrasensitive' attention to its feel. Her ultra-sensitivity can involve (while not necessarily needing to do so) an embodied site visit. Often she does not 'sit in' physically but enters via the imagination; Google Earth, Street View, Instagram hash-tagged with the location of interest, Pinterest

Image 6.1 Urban designer in front of her monitor with one hand on the keyboard and one hand on the mouse. In the front to the right lies an 'atmosphere sheet' on the table.

Photo: Anette Stenslund

or other social media help boost her imagination. I see these early exercises that help the designer to home in on appealing characteristics and features of the site as a kind of *aesthetic screening*. Guided by the appreciation of *wabi sabi* beauty, the SLA designers ask themselves what features are already given by the site that can be re-used, highlighted and emphasised further in the redevelopment plan. The designers screen the landscapes for objects, artefacts, architectural and non-architectural elements like staircases, kerbstones, walls, windows or elements in the outdoor environment like trees, grasses and leaves in order to stage a *wabi sabi* nature feel that appears 'wild' and 'untamed' – stronger than humans. Decaying buildings or artefacts with worn, weathered and rustic appearances are therefore likely to be included in the new landscape design to support the feeling of 'nature's' supremacy.

What is noteworthy, however, is not so much the style or kind of things that catches Selma's and her colleagues' attention (they are described in Chapters 3, 5, 7 and 8) as how Selma attends to them so closely, since only a few would notice them during a busy working day walking through the area, for instance. I am interested here in the way she looks at the subjects of interest and how her perspective, given by the scale, plays a role. Selma homes in on small things in close-up. Sometimes she pays attention to microscopic, tiny matters, crucial to the atmosphere. It is this upscaled perspective that she understands as the 'ultrasensitive' part of her work.

In the following, I refer to an upscaled perspective when Selma zooms in on parts of images in order to relate to them in detail. My mention of downscaled perspectives refers to a reduced scale that happens when designers zoom out.

Materiality in the sense of 'stoflighed'

Selma tells me that she zooms in on features on site that she notices in particular. If she has pictures from the area, she will often start to cut and paste upscaled sections of photographic material into the atmosphere sheet. The upscaled sections on the sheet resemble a patchwork of sub-elements as if they were looked at through a magnifying glass. She ends up with a sheet that holds a selection of constituents that she considers of crucial importance to the atmosphere that she and her team wish to accentuate, develop or expand further through their design. The constituents are usually discussed and evaluated collaboratively.

What comes to Selma's attention through her zooming activity is less about the descriptive or objective nature of the elements that she finds of interest than 'what they can do and should be able to do' with people. The decisive factor in zooming is therefore *how* selected constituents *behave*, that is, what they *do* rather than what they *are*. For example, how a waterfall falls or showers (see also Chapter 7); how trees are blowing in the wind – how they behave in high or low temperatures and whether they change colour or shed their leaves; and how different kinds of road surfaces might reflect sunlight, accumulate heat or treat people walking on it – whether they mute the sound of footsteps or throw them back in reverberation. Material's 'doing' Selma describes as *stoflighed*, and as I ask her what she means by the term, she explains that this is the atmospheric *virkning* (in English 'effect' or 'impact') evoked by constituents within a given space.

In Scandinavian languages and in German, *stoflighed* (Danish) or *Stofflichkeit* (German) refers to the *virkning* (Danish) or *Wirkung* (German) meaning the effect or impact of *stof* that in English can mean both material, fabric, substance, matter, subject-matter and even a drug. It was the Danish architect Carl Petersen who in 1919 introduced the term *stoflighed* to a Danish audience, and in a speech he refers to *stoflighed* and *stofvirkning* interchangeably (Petersen, 1918). The term is reminiscent of Böhme's definition of things' *ecstasy* defined as the way things draw attention to themselves and make their presence felt by 'stepping out of themselves' (2001: 131). Although both *stoflighed* and *ecstasy* refer to approximately the same phenomenon – namely, that matter (human and non-human, biotic and abiotic) transcends what conventionally is considered its own perimeter – there is perhaps a nuance in that the concept of ecstasy notes the simple fact that it is happening, while the concept of *stoflighed* points to the qualitative perception of what radiates from material bodies. However, we are not going to dwell on trifles here. More importantly, it is convenient to know that when I choose to refer exclusively to *stoflighed* (translated into English as *materiality*), it is done out of respect for my informants' own choice of wording. Hence, my understanding of materiality refers to *stoflighed*, not considered as the 'property' of materials, but rather materiality arises in resonance *between* mutually involved parties (objects and subjects) framed by human perception. In other words, materials are no longer 'just' materials, they evoke feelings due to their sensory qualities in experience – for example, how they smell, how heavy or light they are, their surface texture, sound (see Image 6.2).

Image 6.2 Shelf with selected materials stored in SLA's 'model space' used for building, testing and touching materials – among other things, to understand their materiality.

Photo: Anette Stenslund

Below, I pursue how materiality is a sensed interrelation that is both situated in Selma's imagination and in the design process that involves scaling.

Scaling up, zooming in on materiality

As soon as Selma brings together on the atmosphere sheet selected upscaled sub-elements found within an area as a whole, she starts to get a sense of the character of the space. She exemplifies:

> Most people think of Christianshavn as a lovely spot. It's nice. People like it. But if you just take a look at one of the streets, it can be hard to understand why it's nice. So say I were to make an atmosphere palette of the same experience: I would do some zooms of the space and use them to explain why Christianshavn is a nice space to be in. For example, I'd zoom in on the cobblestone paving. It's a natural paving and it adds history, some age, to the space. There's a lot of information in the paving which tells us something about the place and makes it special. And then there are also some large old trees, I think they're chestnut or perhaps ash trees. In any case, they have large leaves that give some volume and character. The trees face a mish-mash of old and new buildings, which brings many colours to the space. This would be my observation of selected sub-elements that make the space feel like Christianshavn, and that allow everyone to recognise it. As an architect, you learn to understand space by noticing such details.

In the above extract, Selma tells me how she captures the atmosphere of a street in one of Copenhagen's highly sought-after dock districts. Zooming in on a few cobblestones, some large leaves on old trees and the varicoloured road frontages of old and new buildings, she extracts what will go into her atmosphere sheet. At this stage, she zooms in – never out – in order to catch the upscaled close-up of the atmospheric constituents of this place. Her view is decisive. Ideally, by looking at the upscaled parts depicted by the atmosphere sheet, a viewer should feel able to almost touch the surfaces of the cobblestones; sense their unpolished raw textures and even identify the lodes in the bricks. Also, the upscaled close-ups should enable the viewer to sense age comprised, for instance, in the solid leaf stems grown by the trees to enable the leaves to hold on to the morning dew, which, as the zoom reveals, accumulates in the form of pearls of water on the leaf surface. And when Selma zooms in on the street frontage of buildings, one should feel able to connect with the crooked and buckled shape of the outer walls with fissures and cracks – some patched and repaired and others remaining like open wounds in contrast to the straight lines of new construction.

Building on visual anthropologist Christina Grasseni's concept of 'skilled vision', which suggests that vision, like other senses, should be considered as an embodied, skilled and trained sense (Grasseni, 2004: 41), I tend to think of the atmospheric scaling that I observe at this stage in the design process as a practice involving a designer's skilled way of screening an area for elements of sought-after aesthetic value, informed in this case by *wabi sabi*. The designer's skilled ways of looking at urban spaces under redevelopment through practices of scaling in digital modelling can, as we are to see, take diverse forms – some more institutionalised than others. When I suggest in Chapter 3 the *wabi sabi* rules of thumb – appreciating nature's grandiosity, celebrating the beauty of things' untidy, unruly, unpolished, transitory, imperfect perfection – it is undeniably me as a researcher imposing an institutionalisation of SLA's aesthetic practice. They do not themselves formulate rules in text; they do not define the Nordic *wabi sabi* approach; they only each speak independently about common principles of their practices. Thus, when all these independent voices are gathered together, a pattern emerges which enables me to suggest how the SLA designer's way of perceiving and approaching the design is often guided by implicit aesthetic principles that I call 'Nordic *wabi sabi*'. In this chapter we see how this might also be the case when Selma screens for materiality that calls for the upscaled perspective for it to make a viewer feel connected to the spatial atmosphere.

Selma demonstrates ordinary skills in composing an alluring image as she applies her zooms to the atmosphere sheet. Any image maker will know how to depict subjects from the point of view of a person interacting close up to them. The involved view makes viewers feel connected, but interestingly the atmosphere sheet is not made for an audience outside the studio. The upscaled parts that Selma selects and brings together on her sheet are meant to guide the team – herself and her colleagues – involved in the project. Hence, the preparation of the atmosphere sheet serves more than the production of atmosphere for future urban spaces, as it also strikes a mood within the team. This charts the direction for how the team are expected to align their design activity in ways that complement and support the atmosphere suggested by the sheet. What I see in situations like these is that the atmosphere sheet acts as a tool that ensures continuity of work and gives a nudge if some drawings start to move away from the original wishes for the design.

It is possible that the atmosphere sheets have served their time at the stage when they ensure that the designers keep in step with each other by connecting to ambitions of an overall atmosphere. A few observations, however, suggest potential transformations and reworkings of the upscaled atmosphere that travel to a downscaled perspective, drawing a link between atmospheric parts and their visualised whole. I follow one of these essential observations below.

Zooming in and out; scaling up and down – ways of bridging parts and the whole

In the office, Anton is working on visualisations for an inlet in the inner Oslo fjord, and as I walk past his desk I casually ask him how he is doing with the visualisations. He looks up from his monitor, rolls out his chair slightly from under his desk and invites me to take a look at his screen. As I step closer in order to study his visualisation (a rendering), he starts to scroll the mouse in and out, in and out, over and over again. I remember how I found his way of constantly changing the picture by scrolling in and out from upscaled to downscaled to upscaled perspective very annoying and disturbing. It prevented me from focusing on what he had drawn in detail, and since I could not fasten on what I thought he wanted me to see, I remember that I was just about to ask him to stop constantly changing the picture. But I didn't. Instead, I came to terms with my confusion, and only later it dawned on me that perhaps Anton did not mean to annoy. He might not have wanted me to cling to one single element in isolation nor in close-up, which was what I was in fact trying to do and what I expected he wanted me to do. In line with Yaneva's studies on scale modelling (2005), where she learns how designers who jump back and forth between large- and small-scale models produce knowledge that in the end leads to the realisation of buildings, I recognise a similar epistemological contribution of movements between scales of different sizes, although here not in scale modelling but in digital modelling. With the mouse, Anton helped me see something other than I had expected. He drew attention to the *link* between the parts in the image and their relation to the context as a whole. It was this relationship between the upscaled parts (of which one would perhaps be able catch a slight glimpse when Anton scrolled one way on his mouse) and the downscaled whole – the context that would only appear on the screen if he scrolled to a zoomed-out perspective – that was of importance.

Just as Grasseni's (2004) ethnographic study of cattle breeders required her to learn to look at cattle in ways breeders would, at this stage I had to learn to look at visualisations like an urban designer in order to access their scaling practices. What I see as I attend many designers' image-making processes is how visualisations, produced for an audience outside the studio, are composed according to the classic rules of composition used in fine arts. For example Margit, an architect, tells me how she always applies the golden triangle in order to exude harmony and balance. It is only when she later inserts a selection of aesthetic units (patinated 'Berlin-like café benches', large trees throwing playful shadows or wild grasses) that she allows herself to play around with the format in order to 'inject some vitality', but it is still her job to ensure consistency between parts and the whole. If large leaves are to retain the dew in the form of pearls of water on the leaf surface, the season, the weather and the time of day need to match the impression, for example with a hazy autumn morning, which will again influence the colour toning of the image, what clothes the cut out people are wearing, etc. One way to check the harmony and how parts fit together goes through a constant change of scale that happens via repetitive movements with the mouse scrolling in and out again and again.

Atmospheric scaling is, as we have seen so far, about screening, selecting and highlighting by zooming in on sub-elements of a given environment that, according to the architects' and designers' aesthetic judgement of taste (a *wabi sabi* taste), have the potential to appeal atmospherically in the studio and beyond. For designers working with materiality (as *stoflighed*) at an upscaled level, the focus is mainly on atmosphere in the design process and the connection and common course of designers within given projects. Scaling up and down for the production of future urban atmosphere has a main focus on the external, client's and users' connection to the image that seeks to evoke a 'digital atmosphere' of future urban space (term from Degen et al., 2017: 9). This latter take on atmospheric scaling as a way of sensing how well or not things come together in an image of a space – and how it can 'inject some vitality', as Margit tells me – shares common ground with architect Frank Orr's understanding of scale (1985). According to Orr, scale is experienced at an intuitive level as a feeling of 'fitness' (or 'un-fitness') of the diverse elements that conjoin in the construction of buildings. Fitness 'conveys the idea of balance, of harmony, of dynamic symmetry, of honest expression of the size of structural elements and, in general, of a pleasing and satisfying wholenesss' (1985: 9). What attracts or repels is determined by how scale aligns or disharmonises with cultural values of beauty or, to put it simply, what a culture finds appealing (1985: 9). Although Orr is not occupied with image-making but mainly focuses on realised architecture, his consideration of scale is phenomenologically driven and therefore it resonates with the atmosphere approach in this book.

Considering scale from an experiential point of view enables one to observe how scale is not only an objectively measurable or visually observable occupation or practice but also involves emotions, aesthetics and cultural values that are not to be deemed a subjective affair of minor importance (Yaneva, 2005: 870). When scaling designers scroll their mice, I see how they are culturally informed by matters of taste and aesthetics of how things go together. It is the experienced designers with SLA DNA who will intuitively know how to compose images that manage to inject some vitality into the orthodox and classic aesthetic rules of thumb in ways that 'disturb' the symmetry, balance and harmony in an elegant *wabi sabi* way. In the concluding section I will describe an employee who suggests how intuition might serve as a quality stamp of the work done by SLA.

Scaling atmosphere by intuition

'Intuition is key', Jonathan, a communication officer in SLA, tells me in an interview. 'This is where art comes in as a part of what we do', he adds. He describes how this artistic nerve in SLA lends their projects an aesthetic surplus – something 'sublime', he says – which, according to SLA's foundation in complementarity theory (described in Chapter 3), does not depend on rational studies but complements them. In architecture nowadays, intuition has received a 'strangely downgraded' reputation which, according to Jonathan, happened as a result of a 'wave of rationality' taking over the architectural world with a few big companies at the forefront. Jonathan credits 'the rational wave' with introducing standard requirements for design companies to be able to explain their choices of design. As a result, urban design no longer finds legitimacy with reference to 'pure magic', Jonathan explains, 'or some divine magnitude' that, according to the stereotype, turns out to be 'an architect – typically an older man', Jonathan smiles wryly, who during a fancy cocktail party quickly sketches an idea on his cocktail napkin, and then this sketch becomes the leading design idea with no further justification needed. Jonathan is relieved that this way of designing no longer finds support in the industry, but also he sees a problem if urban design tips into the opposite extreme 'building everything around rational arguments where intuition is absent'.

The kind of intuition that Jonathan considers 'key' is not the kind that comes out of the blue to a designer, but a qualified form of intuition. This sought-after intuition is, to him, an artistic kind of intuition that he hopes SLA will become better at strengthening in the future. Preferably, Jonathan says, artistic intuition should be unleashed in close correlation with facts, and it requires the designer to bodily engage in the space under redevelopment:

> Artistic intuition is, as I see it, something you can unleash or set free in close correlation with knowledge of your subject field. Intuition completely from out of nothing – that is, when you know nothing about why you do what you do, you just sense it's right –that may be all well and good, it may become a success, but then it is more by chance than intellectual judgement.

When Jonathan considers SLA's market positioning, the production of more and better diagrams and evidence-based documentation is not the right way to go. 'Of course we need to collect a lot of information and show that we are researching', he says, but in order not to stand just on one leg but set off from a position where feet find solid ground, they will need to strengthen their 'premium' by taking a leap into a design approach that cannot be foretold or predicted by way of calculation. This leap involves finding a solid footing in 'poetry', Jonathan explains, a poetry of urban design which feeds through creativity and artistic ability that all comes under intuition. The only reason why Jonathan seeks to avoid the term 'intuition', he explains, is because it evokes the wrong associations of ideas, such as coming out of the blue to be sketched on cocktail napkins, which is not beneficial in terms of sales.

> Today, it is expected that everyone will be able to document their value creation. For example, if you say on a project that you can extend the summer season of an area, then you must also be able to document that you actually know it, and the argument will then be 'we extend by three weeks because we can plant these trees in this way, which takes 93% of

all NOx particles because the trees clean the air, and we can actually go so far as to say that the body's stress symptoms are lowered by 8% if you go through this green passage' and so on and so forth.

The scientifically rational foundation is the one leg, to stay with Jonathan's metaphorical wording, that nobody can do without, but he also believes that it is necessary to take seriously that architecture, landscape architecture and urban design in general are an art form. 'It is one of the bound art forms, but it is an art form', he says, and for the studio, this means that they see a call to find a position where aesthetics meets rational argumentation in 'poetic solutions'. I ask for examples of projects where Jonathan believes there is poetry, and he says:

> A place has acquired a degree of poetry when it is experienced by its materiality [*stoflighed*]. The materiality in what we encounter, that is the poetic solution. And it's hard to sell, show or talk about this before you go there and place yourself in the middle of it. Then you will discover it. So it's a challenge [in terms of marketing].

Poetic (intuitive) design solutions are, according to Jonathan, driven by embodied experiences of places on site:

> You cannot form opinions about atmospheres as the wind blows. [...] This strategy is just as incomplete as a rational, scientific or diagrammatic strategy might be, because both are lacking the embedded knowledge of how a place might be.

It is the interweaving of rational and artistic arguments that he sees as the future market strategy of SLA – one that is not content with resting on subjective sensations or on objective scientific standards. Both camps are too far from the empirical reality that is to be experienced and felt, he says.

Atmospheres are corporeally experienced, and I cannot think of one ethnographic researcher who would disagree with Jonathan, emphasising the advantage of engaging bodily in subjects that one would like to know better. No doubt designers can benefit from bodily experience of urban spaces. During the fieldwork, I participate in site visits with SLA designers, but I also see – as in this chapter – how employees are often urged to orient themselves using imagery and their own imagination when budgets do not allow the designers to go on field trips. Two things are, however, worth noting about SLA's work, for instance in scaling. Firstly, the work does not happen 'out of the blue' as a call sent from nowhere or from heaven. When they scale up and down the materiality paying attention to the bigger picture and its atmosphere, it happens with an activated cultural skill, where they draw on personal experiences of what it feels like to be in similar surroundings with a comparable materiality. This scaling task demands their cultural skill in empathically imagining the felt resemblance of various materials, which can be advanced in their design. Additionally, if scaling serves to 'inject vitality' to design projects, it requires the skill of an experienced SLA designer with SLA DNA – one who has developed a sense of the good taste (read: Nordic *wabi sabi*) desired by the studio and nurtured the cityscape.

Conclusion

Architectural scaling is a quantitative discipline in geometry and proportionality. The chapter suggests, however, that digital scaling is also about atmosphere in imagery and as a workplace design supporting colleagues' ability to cooperate. Hence, scaling as space dimensioning is not only responsible for the practical, functional and quantifiable effect of spatial design, it also has to do with the emotional experience of space and how to make one's peers tap into the felt dimensions of spatial design. One way to focus on the atmospheres of urban design is to zoom in on the materiality of constituents of a place depicted from the 'involved' and 'engaged' perspective in order to make all employees recall the same emotional quality that is to lay the foundation for a project. The chapter looks at observations that may suggest how digital scaling is also about bridging the felt qualities of sub-elements with the overall composition and unified visualised whole. In SLA, this bridging is preferably not idiosyncratic but guided by a kind of artistic intuition based on skilled bodily experience.

References

Böhme G (2001) *Aisthetik: Vorlesungen über Ästhetik als allgemeine Wahrnehmungslehre.* Munich: Wilhelm Fink.
Degen M, Melhuish C and Rose G (2017) Producing place atmospheres digitally: Architecture, digital visualisation practices and the experience economy. *Journal of Consumer Culture* 17(1): 3–24.
Grasseni C (2004) Skilled vision: An apprenticeship in breeding aesthetics. *Social Anthropology* 12(1): 41–55.
Ingold I (2007) Earth, sky, wind, and weather. *Journal of the Royal Anthropological Institute* 13: 19–38.
Orr F (1985) *Scale in Architecture.* New York: Van Nostrand Reinhold.
Petersen C (1918) *Stoflige virkninger: Foredrag holdt på Akademiet i Februar 1919.*
Stavric M, Sidanin P and Tepavcevic B (2013) *Architectural Scale Models in the Digital Age: Design, Representation and Manufacturing.* Vienna: Springer.
Yaneva A (2005) Scaling up and down: Extraction trials in architectural design. *Social Studies of Science* 35(6): 867–894.

7 ASTONISHING ATMOSPHERES
MIMETIC DESIGN

This chapter argues how a mimetic design, from the perspective of SLA designers, creates surprising atmospheres. Via four selected examples discussing a waterfall's fall, a frog's croaking, a corny monkey cave and a stone paving's calculated randomness, the chapter identifies general *ways* of designing urban nature in SLA that support a particular type of astonishing atmosphere. In the conceptual analysis, I borrow from Michael Taussig's explorations of the mimetic faculty having just as much to do with copying, imitating and making models as it has to do with alterity, creating difference to 'become and behave like something else' (Benjamin cited in Taussig, 2018 [1993]: 15). The modes of mimicking activities that I identify through conversations with employees span from photographic fidelity to imperfect yet effective copying and to dogmatic assumptions of a more speculative nature.

Ideal typification of astonishing atmospheres

There are many ways to express how SLA's nature-based design should ideally land in experience. Nature's greatness can be perceived by the designer in marvellous epiphany, wonder and astonishment, it can astound, amaze and allure in a variety of sensations that I will conceptually group together as an ideal type of astonishing atmospheres. I therefore do what Andreas Rauh explicitly does not do when he defines astounding atmospheres not in terms of typification but as atmosphere that turns one's perception into the very object of perception at a conscious level (2018: 13). I am not arguing that astonishing atmospheres can sharpen self-reflection around perception's own atmospheric power, nor am I arguing against that suggestion, but since in the ethnographic work I am faced with a reality that has a much narrower understanding of atmosphere than the one I conceptually subscribe to, there is a point in clarifying this difference, and the ideal type will serve this purpose.

Until now, I have referred to atmosphere as the most fundamental and inescapable way to be involved and embedded in one's surroundings. Atmospheres are spatial feelings that float in the air and are constantly reshaped due to the situations that frame them. Atmospheres are therefore not anchored to the ground in terms of locations, not bound to objects or human beings, but rather they emerge from situations where humans and

DOI: 10.4324/9781003279846-7

physical entities (material and immaterial) meet and become involved. Human beings are always situated and feel atmospheres whether these are boring, trivial, awfully exciting or astonishing. This understanding, however, is different from the one I meet in the studio. In SLA the urban design either evokes atmosphere or it does not. A telling example is given by a group of designers who shake their heads as they hear about my interest in following a project that, according to them, is 'absolutely devoid of atmosphere'. Atmosphere is a treasured and sought-after quality in the sense of an added amenity value. Poor quality design does simply not have it, from the designer's point of view. This much narrower conceptualisation of atmosphere is tied to a Kantian understanding of aesthetics, defined as what stands out as 'tasteful' according to its beauty. Hence, if atmosphere is defined as something that appears in aesthetic engagement, and if aesthetics is tied to cultural norms about beauty instead of – as in my definition (see Chapter 1) – being related to common sensory felt perception, atmospheres consequently reveal themselves only in the special cases where the surroundings move us by their beauty. In the studio, as the reader will know by now, beauty does not follow Western orthodox opinions about beauty, but rather Japanese aesthetics defined by the *wabi sabi* beauty of imperfection. With my ideal typification of astounding atmospheres, I am able to analytically address what SLA envisions as atmosphere at all – found in the *wabi sabi* beauty of the nature-based design.

Below I analyse the four selected examples from SLA's design practice that serve to illustrate how mimetic design techniques, some of which call for experiential uncertainty, can evoke astonishing atmospheres.

The waterfall – how water falls

In an interview in which Mattis, himself an architect by training, describes SLA's approach to me by distinguishing it from conventional 'tight classical architecture', he tells me how nature-based design transforms the urban space he currently works with in the city of Stockholm. The transformation is enabled by the 'form language of materiality' (see Chapter 6 for an elaboration of materiality) that he introduces to the site, characterised by 'soft transitions' that break with the existing grid-like straight lines that he observes dominate the area. He wants to avoid, he says, that 'it gets too tight geometrically. It should look like it's random. It must be something you want to touch. Something you're drawn into and get curious about'. Mattis then shows me four images of waterfalls that he uses for his suggestive work. He says:

> This picture [image 7.1, option 2 in the middle] shows that there is unpredictability in the way the water behaves. The surface may just be rippled, but the water runs unpredictably, presumably because it's affected by wind and weather and by the forces that only nature creates. This is often what is quite interesting: to take something that is rational – that is, take a surface, a pattern, a structure or a grid, which is logically composed and calculated – and then you expose it to natural influences, and then something unpredictable happens.

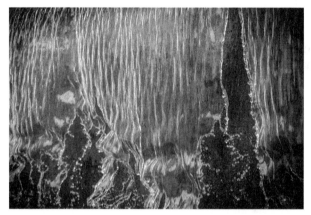

Images 7.1, 7.2 and 7.3 Examples of water falling. Option 1 (7.1 top) shows large
  stones, and water that showers as it finds its way between
  them. Option 2 (7.2 middle) shows Mattis' favourite of an
  unpredictable waterfall with stray patterns in a solid water
  surface. Option 3 (7.3 bottom) is what Mattis calls a predictable 'water curtain' due to the water falling heavily down
  unaffected by wind and weather.

Unpredictability here proves vital for the creation of atmospheres that appeal, and Mattis elaborates about why this might be the case. He starts to compare the unpredictable waterfall that he enjoys (7.2, in the middle) with yet another waterfall (7.3, bottom) that he enjoys less. In comparison to the interesting unpredictable waterfall (7.2, in the middle), the alternative, to him, is just a 'boring water curtain', and he justifies with a double negation: 'because it's not unpredictable'. The water curtain (7.3, bottom) apparently suffers from a predictability in the way its water falls, and in experience predictability leaves no space for things to surprise. Astonishment, it seems here, requires some degree of uncertainty about how the surroundings will behave and how situations will unfold. Predictable behaviour, by way of contrast, turns a waterfall into something boring. This is not the result that Mattis aims for. The water curtain has 'a static expression, and I'm almost certain that it's designed by a building architect', he giggles, as if to indicate that architects are static by nature. 'It may well be', he goes on, 'that economic, political or practical constraints and obstacles will cause us, in SLA, to make a similar massive, large and roaring water curtain', but most important at this stage in the process is for Mattis to explore how the water may fall in exciting ways. And he explains that he is interested in how it can appear in a more materially unpredictable way, which according to him is the atmospheric characteristic that is at the centre of SLA's way of working.

Based on a broad conceptualisation of atmosphere, described above, one might argue that predictability has as much atmospheric effect as unpredictability does. Both appearances are atmospheric: one might be less exciting than the other, but both ways of perceiving the waterfalls are atmospheric – they touch us mood-wise and mark our being. With a narrower conceptualisation of atmosphere, however, only the unpredictable waterfall is atmospheric. Due to its astonishing way of behaving in ever-changing ways, it leaves us with a hint of uncertainty in how to engage with it, and therefore it becomes attractive as a design solution. The unpredictable waterfall offers an experiential uncertainty that is also given in many magnificent nature experiences. Mattis says about the attractive waterfall solution: 'It's more stimulating and interesting. It makes you wonder, and it makes you want to get closer, and it forces you to be present' in a different way than is the case with the roaring water curtain, which is continuously the same. The unpredictable waterfall attracts your attention because it pulls you out of your common-sense everyday interaction with the world – it intrudes, which gives a different experience than if one had voluntarily turned one's attention towards it (see also Alfred Schütz's imposed and voluntary criteria of relevance, 1975: 190). In this way, the unpredictable waterfall does not allow habituation, and the reason is its multisensory nature: it constantly changes its character not only for the eye, but also for the way it touches, metaphorically and literally, by splashing and sprinkling at uneven unpredictable intervals, and, for the ear, it contributes to relentless shifts of the sound-scape. Hence, the desired waterfall attracts attention by its resistance to behave in expected ways always holding the potential to be different – out of human control. The water curtain, on the other hand, is the same all the time, and although it may rumble with much higher sound frequency, it can still subside for the ear because sensory adaptation tends to rule out con-stant stimuli – just like when those living near a railway station no longer hear the trains pass, or when perfume wearers are no longer able to register their over-deodorising.

But are 'nature' and 'the aesthetics of nature' always exciting? Can nature not be boring? And is water falling due to gravity not equally natural, regard-less of whether it falls one way or the other and whatever pattern it makes on the water surface? Of interest to me here is not the social construction of nature/culture-relativity or the nature-philosophy on which SLA rests its business; what attracts my curiosity is how form and atmosphere relate. In the case of the waterfall, it is not indifferent to the designer how the water surface shapes a pattern as a result of its interplay with nature's forces of gravity, wind and weather and the rocky, uneven surface it runs across. The water falls naturally whatever the resonating relationship is, I would argue, but apparently some shapes and patterns created on the water surface seem to be more atmospherically attractive than others – the water curtain is bor-ing and lacks atmosphere in comparison to the exciting unpredictable one that manages to surprise in experience. Form in terms of shapes and patterns is therefore crucial to the atmosphere of the design, but for SLA only some forms and patterns (whether man-made or created by forces of nature) are desired. The unpredictable forms, that take shape in ways that at least appear to be beyond the full control of human beings, are the ones that appeal to the SLA designer because they resemble the idea of nature's grandiosity.

SLA's city-nature is a designed and man-made nature, and therefore the attractive and thus atmospheric patterns of nature can be calculated and constructed. The tricky task of the skilled designer is to calculate ways that make calculation disappear in the experienced end result. Calculations needed for construction work should only serve as scaffolding. In other words, the city should not feel constructed, it should feel the opposite: untouched by human hand within the spaces under SLA's guidance. As the waterfall case demonstrates, however, a clear division between built and grown, touched and untouched seems impossible. Water falls 'naturally' due to gravity whether the ground surface it runs over is designed or not. The water may well fall differently depending on different surfaces, but it falls equally naturally. Thus, the appearance is completely reversible and detached from any degree of naturalness. There are, therefore, no longer any ups and downs in the nature-culture distinction and the section below pursues how the nature-based design can mime nature in ways that produce astonishment by their artificiality.

The frog's croaking

In 2001 SLA was given the task of turning Frederiksberg City Centre, a centre and a commune encircled by Copenhagen City, into a liveable urban space that would offer more than just a place for transit. SLA responded to the task by introducing, for example, water elements, specific lighting features and some playful sound effects. I will fasten on the soundtrack of a croaking frog played with the intention of creating a nature experience for the city dweller and at the same time arousing curiosity and thus creating astonishing atmospheres. Walking through the city centre that links a metro station, high school, business school, library and shopping mall, some will occasionally hear a frog. Apparently not everyone hears it, but at an internal meeting Stig L. Andersson tells his colleagues an anecdote about a little girl who passes through the city centre together with her mother. It is the girl who hears the frog, upon which she tugs at her mother's hand and asks her to stop. The girl's excitement nurtures her desire to start investigating where the croaking comes from, and the girl and the frog together turn what was mainly serving as a transit area into a playground (Image 7.4). The space becomes a place where people choose to stay for a while or simply stop briefly before rushing on. But it is not just what people do and how they

Image 7.4 Girl playing at Frederiksberg City Centre with grownups passing in the background.

Photo: SLA

behave that is of interest to SLA. Of interest to the designers are people's experiences. What Andersson tells his colleagues about the girl's frog experience testifies to SLA's focus on experience:

> Whether or not the experience is awakened by natural or artificial frogs, it is real. Experiences can't be fake. The girl's nature experience is not any less authentic or intense just because the sound of the frog is staged and not caused by a real frog. It's just different. Neither greater nor lesser.

It is not that you should necessarily feel like you are walking through a bog when you visit or pass the Frederiksberg City Centre. Few people would imagine themselves in a bog or swamp somewhere out of the city. Of interest, however, is what sound does to people and how sound affects people's mood as they engage with it. Different people in different situations engage differently with their surroundings. In Andersson's anecdote, the child is vigilant, and unlike the mother and other grownups, does not rush through the area shutting herself off a priori from her surroundings. The sound of a frog croaking arouses curiosity because suddenly you need to know where the croaking comes from. The frog instantly invites the girl to play hide and seek; she starts to search for it with the same kind of curiosity as she would probably have felt in 'nature', in the countryside in a bog somewhere. The adult, her mother, is invited to join the game with the same kind of fascination and excitement and perhaps she will. Her mother might not expect to find a real frog, but she might become fascinated by the unfamiliar situation of the sound of a frog 'out of place' and she is likely to be driven by curiosity to unravel the mystery of who created the frog, how and from where it is played. The astonishing atmosphere derives its force for drawing in passers-by through elements of uncertainty and unpredictability as to what, where and why the croak of a frog is present in a city centre.

How users actually experience the space is something only they will know. This is not what occupies my interest. I am interested rather in the design concept described by SLA, their intentions and in this case their mimetic flair. The sound that is included in the design imitates the croak of a frog. In experience it is to appear *as* a croaking frog, and this way of using imitation in the design strategy grants the reproduction of a croaking frog the character and power of the original, while at the same time seeking not to be identical to what it imitates (the frog) – it appears as something else (alterity), namely, a recording of a frog installed at a place in the middle of the city, which is how the design yields its otherness. Unlike the example above with the waterfall which should not appear as a design in the end, the composition of the frog design should in turn not be obscured. The design, in other words, does not erase itself, rather it draws attention to itself by being open about its artificial construct and thus it appears honest and authentic in experience, even if its effect is sensory deception. Rather than evading attention, the design draws attention to itself; in mimesis it insists on its alterity.

In the third example I describe a design solution that fails in the mimetic dialectic and thus ends up becoming corny or simply ridiculous in experience according to the designer – with no astonishing atmosphere from the designer's point of view.

The monkey cave

Unlike the two previous cases, the following is not a project by SLA but an illustrative example of a design approach which some SLA designers fear they could end up following. 'If it happens', Gudrun, an experienced architect, tells me, 'we need to correct it'. She shares with me her pet hate in terms of a bad design solution devoid of any kind of atmosphere (to her): the monkey cave. I will refer to the monkey cave as a telling example of a mimetic design approach that loses the ability to distance itself from the reference that inspired the design. With the loss of mimetic dialectic in the design approach comes, in this case, a risk of the design ending up being corny in experience.

In an interview, Gudrun tells me how the same experienced type of atmosphere can be evoked by different design. She goes on to talk about a 'delicate balance' that she often encounters in the design process where she senses a risk of the design 'tipping over' to become 'corny'. This can happen, she says, when the design tries to look like something it is not. In an attempt to have her expand on what exactly she means, I ask when she last experienced this risk, and she says: 'I can give you an example', and then proceeds with a question: 'I don't know if you know the rock formations – I think they are in Iceland – that look like hexagons and stand in pillars?' Gudrun is thinking of the basalt pillars created by a dark fine-grained type of lava. The pillars are found in both Iceland and Scotland and their hexagonal shape is due to a late cooling that has caused them to crystallise before they have solidified. 'The wonders of nature are unimaginable, and it's hard to believe they weren't made by humans', Gudrun says.

A little investigation on my own informs me that the basalt columns have provided fertile ground for pagan and fabled tales of having been created by the gods or otherworldly forces. They suggest a truly astonishing appearance, Gudrun argues: 'People always get like "oh wow" when they see that [the pillars] are created by forces of nature. It's quite amazing'. She then explains that the pillars hold a quality in experience that is highly attractive to her and her colleagues, but Gudrun has noticed a dead end in her own work as well as in the work by others, where one begins to design in too much detail in order to achieve the sought-after atmospheres similar to

Image 7.5 The monkey cave in Copenhagen Zoo.

Photo: Mads Søndergaard and Esther Eo

those evoked by magnificent natural phenomena. The design of the monkey cave is her pet hate (Image 7.5). Gudrun explains:

> Using the basalt columns as a reference point, one might want to devise a principle for a pillar wall that should create the same sensation as one would feel in nature, and this is where I sometimes get a little arrogant and call it a monkey cave trend, because it looks like what you make for monkeys in a zoo. And I'll have to apologise for my example, I am not trying to condemn zoos.

It is not just monkey caves or zoo reserves in general that Gudrun problema-tises. She critically assesses all types of design, including her own failures, that stray into too strong a focus on imitation, leaving little or no room for coinci-dences in the formation process. She can sometimes find herself in what she calls a dead end, drawing too many details in order to control how the design will turn out, but 'this is absurd' she says, there is no chance that a project will be realised on a 1:1 scale – the idea as it looks on paper never becomes reality, which has to do with the situated context dependence of design and atmosphere (Stenslund, 2017). Gudrun speaks about overworking the design when she or her colleagues unwittingly become lost in overly detailed drawings that pursue accuracy, as if to determine the design's outcome instead of leaving it to chance. Gudrun explains to me how a series of scenes in a zoo have been created in detail as a reminder of the animals' original setting:

> It usually looks terrible, and it's super complicated, time-consuming, and expensive to make. To do a monkey cave you would need to draw all the details, and nothing would be left to chance, which is just weird or boring to look at. So, taking a very specific image of nature and saying 'we're just doing the same thing again,' well, that's difficult and it will not give the same feeling at all.

Overworked and overly detailed design is often carried out, according to Gudrun, in situations where the designers attempt to steer projects in very specific ways – for instance when seeking to recreate specific atmospheres of specific (natural) environments. The environments in a zoo are telling for their failed imitation of specific settings with no consideration of the contextual change. The example shows clearly that atmospheres are not tied to materials nor to form alone, but that they are culturally situated and experienced holistically. Thus, hexagonal pillars, no matter how well designed, would never be able to evoke the same astonishing atmosphere as the basalt columns in Iceland and Scotland apparently do. And I will argue that not even if the very same columns were imported to the zoo would they be experienced in the same way. Their atmospheric attractiveness would fade instantly because their experience is not linked to materiality alone – it is situated.

Imitations that aim for exact or approximate copying seem doomed to fail, which leads to another important lesson: designing experiences and atmospheres is not just about designing for users 'out there'; learning to deal with atmosphere requires designers to engage. As I began this book by recalling, atmospheres may be designed, yet no designer is in control of their appearance as they emerge. This necessitates a readiness to work with degrees of uncertainty in the design process itself. When designers seek certainty and control – as under fully imitative efforts in retail design – they are no longer receptive to what comes forth in the design. In such cases, they do not resonate with their drawings and other design activities. But where there is no resonance, there is also no openness to become surprised or astonished by coincidences and the design is at risk of falling flat in experience. At worst, the design is perceived as dishonest or unreliable when it tries to be something it is not, and at best it might just be a bit ridiculous, as in the case of the monkey cave. The result, in any case, is that the atmosphere is different than it was hoped it would be. As mentioned, atmospheres tend to elude any architectural design studio's attempts to invoke *specific* spatial atmospheres, and the monkey cave is an example of this. At the same time, atmospheres likely tease those studios that choose not to make them part of their services, for atmospheres continuously show their presence,

and designers will never get rid of them. If designers do not consider the situated nature of omnipresent atmospheres, they could end up with unwanted
kinds of atmospheres that devalue their design.

Below, in an example calculation for stone paving, we will see how principles of randomness in computation can help avoid overworking the detail
of the design and thus save the designer from the failures of pure imitation.

Stone paving's calculated randomness

Sometimes when Gudrun gets lost and finds herself on the wrong track
in the process of far too detailed designing, she will need to pull herself
out of the specifications and start to think in terms of *principles* instead
of *detail*. As a starting point, she is always working to create some kind
of a random expression. This is the general feature of her design practice
independently of project specifics. If she succeeds in making the design
appear as if it happened by chance – 'as in nature' – it is more likely, in her
wording, to 'attract' and 'fascinate'. Gudrun and her colleagues share with
me independently their countless examples of projects where they feel they
have either succeeded in achieving or failed to achieve the attractive atmospheres that I call astonishing. Listening to their stories I come to realise how
notably the atmospheric allure is created with computer-generated support.
Gudrun takes out a drawing that she has rejected and says:

> This one's not good. Again, we're dealing with a situation where I sud-
> denly ended up making too much detailed design. Well, it's kind of
> like the monkey cave. I suddenly produced too many drawings with
> no great result. The simple drawing, in return, that merely suggested
> using the same [built] element with the same repeating rhythm [...]
> worked much better. The point is that there is no need to draw in
> the variation in detail because I have a principle of randomness that
> solves it for me. The principle makes it look accidental, but of course
> it's not. It's built and calculated. But when I draw, I do not have to
> deal with how exactly the moulds should be placed in order to end up
> making an unpredictable pattern. It happens by itself if my principle
> is good enough. So as a designer, I apply a principle that makes it look
> like it's random, but it's not. It gets cheaper and cheaper the smarter I
> am at making it, and the more dynamic I can make it look, the more
> real it becomes in a way.

A 'good-enough principle' helps Gudrun to reach the atmosphere that she
is after in her design. In addition to the principle of randomness contributing with a lightening of her workload, we see here how it also serves a
crucial function within the very process of creating the right atmosphere. It
is not only the principles that will determine the material design that waits
to meet its user in order to produce atmosphere within the urban landscape;

it also engages with the designer who meets its atmosphere with a curiosity about its development that is no longer given and meticulously prepared by the designer but partly unpredictable to her. I call this way of designing with principles instead of in detail a *mimetic design* based on imperfect copying, for it is not a question of *copying* in detail, but rather of *imitation* in principle. Gudrun says that if you imitate purely, it becomes almost an impossible task to land the project well: 'There are too many parameters that need to be considered and worked out with too high a risk of it ending in failure', she explains to me. It is much safer to formulate some principles for 'the locked parameters' – that is, the conditions or circumstances that the designer is sure to know are mandatory for the design – and from there leave the rest (the details) to a computer. Gudrun talks of her principles as being 'principles of randomness' because they are to ensure the unpredictability that attracts and potentially astonishes. In the quote below we see how these principles are carefully calculated and computer-generated. Frida, Gudrun's colleague, explains how she works with three principles of variation and randomness for a stone pavement:

This paving [image 7.6] is made with a script in Rhino. I formulated a premise asking to mix six different stones in a colour spectrum ranging from very deep dark grey to lighter variations of grey. At meetings with

Image 7.6 Interviewee Frida points to a visualisation of the computer-generated colour graduation of paving.

Photo: Anette Stenslund

the client, it has been decided that there should be light-coloured paving in areas where the road makes transitions [at crossings for instance] and it should mainly be dark within the middle points, which is then my second premise. And then, a modulation towards the bright areas is needed. How do we make a modulation that appears natural? Well, I formulate yet another premise pinpointing where exactly it should be light and where it should be dark, and then finally a script is made in Rhino.

Rhino (Rhinoceros) is a 3D modelling tool that follows the so-called Non-Uniform Rational B-Splines (NURBS) mathematical models that help describe any 3D or 2D with a high level of accuracy and detail. A script in Rhino is the principles or what Frida also calls the locked 'premises' or parameters for the construction that the designer formulates in order for Rhino to do the calculation. In Frida's case the parameters would include the six colours of grey, the measurements of the bricks and the locations of where exactly the road is going to appear light and dark, respectively. The algorithm given by Rhino will then calculate how the bricks can be distributed. Frida tells me that, in theory, and if she had had all the time in the world that she needed, she would not need the computer to make a pattern that looks random. Then, she would have started to calculate a 'grid' herself, as she calls it, that would mix the stones and then she could just explain to the paver afterwards how he should simply obey her few premises, but beyond that, she would invite him to arrange the bricks according to his gut feeling. The problem is partly, however, that it would be time-consuming for her to calculate, and partly that she cannot count on pavers to set aside their schooled understandings of beauty in order to make a random mix of stones. 'It is as if these people have been destroyed by schooled ways of creating symmetry and order, so they quickly fall back on common norms of when something looks nice', Frida says. As we know by now, in SLA they would not necessarily share the *comme il faut* understanding of beauty, and therefore it eases the process, Frida explains, to send a colour-coded computer-generated calculation to the paver and engineers that can help assist their construction work.

This empirical observation that the urban designers use CAD (*computer-aided detection*) to help ensure the experience of astonishing atmospheres that hold an element of the incidental is startling, since a significant proportion of research literature has argued to the contrary. For example, in Tim Ingold's (2007) advocacy for the non-linear, uncertain, curly and disrupting ways of drawing design and architecture, the CAD design is mentioned as a threat that fortifies a late-modern 'workmanship of certainty' where results are 'predetermined before the task is even begun' (Ingold, 2007: 161). CAD introduces 'straight lines' of reasoning – lines of certainty that 'can be specified by numerical value, [and thus] becomes an index of quantitative rather than qualitative knowledge' (Ingold, 2007: 167). Along much the same lines,

Richard Sennett argues that CAD programs 'calculate to the marvel', causing the designers to no longer feel or sense the texture of their design products (2008: 41). The present book and the ethnographic work on which it feeds find crucial inspiration in and great admiration for both sources mentioned. Nevertheless, its findings invite critical dialogue with approaches that mainly bemoan the technological development where flat screens are accused of decomposing human skill and creativity (Ingold, 2007: 127). In the example above, Frida's and Gudrun's calculations may perhaps fortify certainty at one level – they guard against miscalculation and do not leave paving to pavers who have not understood their aesthetic *wabi sabi*-inspired preferences – but simultaneously, the calculations are encouraged by an attitude that embraces uncertainty. The challenges of avoiding detail and only sticking to principles are considerable for the mimetic design that in this case allows alterity through the small gaps that locked premises leave up to algorithmic processing and whose outcome the designers cannot foresee.

Conversations with the urban designers suggest that wonder and astonishment can be rekindled through a computer-facilitated processing of principles, and the analysis suggests how atmosphere in design can be successfully produced by entrusting some parts of the design process to technical aids. The example with stone paving points to the designer's skill in drawing out principles, keeping the overview and outsourcing some decisions to algorithmic patterns as a valuable resource. The designers are prepared to be surprised when they see the algorithmically processed stone pavement visualised on the screen. They do not know the result in advance but are open to what shows up. Frida shows me several pictures of the paving, and I imagine I can feel her fascination; at least she is well pleased with the outcome. 'It nearly shimmers', she says at one point. Who would have thought that computer-calculated cobblestone paving could be shimmering? When something shimmers, it usually breaks the light in so many different ways that it looks like it is moving a bit – the paving also moves in a figurative sense by its astonishing atmosphere.

Conclusion

This chapter has offered a number of analytical reflections of selected empirical cases in order to consider the complexity of a design practice that mediates mimesis and alterity in various ways in order to produce exciting, enchanting, appealing, or what I group together as astonishing atmospheres, in urban design. By exemplifying four instances of mimetic design described through the designers' perspective on the ins and outs of the creation of atmosphere, a picture emerges of a design practice that seems to create a business around the production of astonishing atmospheres.

What appears to sustain the appeal of the design is mimesis that maintains a dialectic between sameness and alterity. The analysis has shown how pure mimesis, as in reproduction, easily ends up in a corny outcome that loses the atmospheric allure. This is where the designer plays an important role in their approach to the work process, attending to uncertainty in various ways, such as when they allow themselves to sense the atmosphere of the design, whether it is natural or artificially produced, reproduced or calculated in origin. In short, they are not able to do the design without experiencing it as atmosphere.

References

Ingold T (2007) *Lines: A Brief History*. London: Routledge.

Rauh A (2018) *Concerning Astonishing Atmospheres: Aisthesis, Aura and Atmospheric Portfolio*. Mimesis International.

Schütz A and Luckmann T (1975) *Strukturen der Lebenwelt*. Neuwied, Darmstadt: Hermann Luchterhand Verlag.

Sennett R (2008) *The Craftsman*. London: Penguin.

Stenslund A (2017) Being in art: A socio-aesthetic study of art in hospitals. In: Manly A, Rønberg L and Jørgensen M (eds) *What does art do in hospitals?* Odder: Narayana Press, pp. 39–59.

Taussig M (2018) *Mimesis and Alterity: A Particular History of the Senses*. London: Routledge.

8 DRAWING ATMOSPHERES
RESONANCE IN DIGITAL GRAPHICS

The idea of 'architectural selves' (Yarrow, 2019: 111), 'designers' (Chumley, 2016: 139), 'artists', 'makers' (Ingold, 2013) and other creative architectural professionals (Saint, 1983) who find themselves at the heart of the process of creation has gradually been provided a valuable counterweight not only by Dana Cuff's demonstration of the collaborative reality of architectural undertakings (Cuff, 1992) but also by Albena Yaneva's thorough Actor Network Theory approach that assigns agency to non-human actors, reminiscent of architects' dependent practices and realisations (Yaneva, 2009). This chapter pursues the same track but from a different theoretical lens that draws inspiration in particular from Tim Ingold's thoughts of correspondence in processes of making (Ingold, 2013). Detailed ethnographic observations will show how encounters between solar/shadow algorithms, curves and structures in construction work resonate with the architect's empathetic readings – and thus how drawing atmospheres for future spaces is an occupation that also requires the designer's readiness to be emotionally drawn into their graphics. Since the conceptual understanding of correspondence may have a material main focus, I suggest expanding the understanding of the corresponding encounters that I observe in phases of design for construction with the concept of resonance. This conceptual development is made for the purpose of envisaging a mutual exchange and transformation of an atmospheric nature where the designer produces drawings while they are themselves drawn in by digital drawings. The chapter suggests how resonating atmosphere in drawings is a way of manoeuvring the uncertainty, picking up on potentials towards future atmospheric urban design.

The case of the quay skirt in Bjørvika

'Let me just start by saying', Sophie stresses to me about a year and a half within the process of working on a specific area in Bjørvika – a neighbourhood in the southeast side of Oslo, Norway, 'this project and our engagement in Bjørvika is a unique affair that allows a lot of drawing. Usually, we don't have time for developing that much through drawing. We normally don't have time to dive in, test and try out options'. Most often, Sophie

DOI: 10.4324/9781003279846-8

says to me, she and her colleagues at SLA need to deliver speedy solutions, but the Bjørvika case is an exception to the rule. From Flora Samuels we know that clients are absolutely key to the incentivisation of the creative and innovative procurements of architectural design (Samuel, 2018: 201), and with a less tight timeframe than is usually the case, this project seems particularly promising for the tenderly crafted result. Sophie is allowed to 'nerd through', as she enthusiastically tells me, on investigations that require time. She explains how the project with its countless subprojects has evolved into a 'rather long process' of collaborations, with many opinions running back and forth in countless email threads between the studio, the client, manufacturers, engineering contractors, etc. 'It is exceptional to be involved in that many negotiations and to constantly be met by wishes that need to be addressed in the drawings', Sophie says.

The extraordinary collaborative team empathy within this project ensures that the investigative drawing activity is brought to the fore and captures my attention as a researcher. As we are to see, the epistemic development of the skirt happens *with* the drawing. Of interest to the discussion here are therefore not so much the negotiations and the potential power struggle between the parties enrolled in the development of Bjørvika as the drawing activity itself and the *way* Sophie draws through a process that prompts her emotional response and bodily engagement in what is revealed through the design. 'We draw to emphasise our points for the client. It is not enough to say what we think is right. We need to illustrate what we mean', Sophie says. But what does drawing mean, I ask? What makes drawing special, I add? This chapter invites Sophie, who is an experienced architect at SLA, to take us on a journey that shows how drawing involves more than simply passing on information and illustrations to a client. Drawing, as she demonstrates, requires the designer's empathic commitment. And all the while Sophie is busy drawing atmosphere in the design, she herself is drawn into the drawing that plays out in resonance between her investigations, the light's reactions, shadows and shapes tuning the space with 'soft' or 'sacral' vibes depending on 'the buttons she pushes'. Perception of atmospheres is to be understood as a kind of resonance phenomenon, Böhme reminds us (2001: 47), but little is yet known about resonating atmospheres as they play out in practice. The ethnographic study presented in this chapter draws inspiration from Tim Ingold's considerations on material correspondences in design. It helps to attend to the ways that the designer is *with* and *among* the units that are part of the design, rather than above or beyond them (Ingold, 2013: 69).

The data I draw on is taken from three interviews with Sophie and countless informal conversations that I have had with her and her colleagues. In the office I watched Sophie for hours, captivated by images on her screen which, for the most part, happened in silence and with tiny movements of the mouse interrupted only occasionally by colleagues' small talk. Included are also observations from internal team meetings that provided me with the opportunity to learn from the way the Bjørvika team discussed their drawings. First, a bit more information about the project and its context.

Storytelling seized by nature-based design

This chapter takes the drawing of cladding for a sheet pile wall as its focal point. From here on I will simply refer to the wall as a 'skirt', as it is colloquially called. In architecture a skirt is basically a board that runs along the border between a vertical and a horizontal surface. It can be between a house wall and the floor, or, as in this case, along an earthen wall and the bed of a river. The skirt is made for Bispekilen Øst – the eastern part of a wedge in the urban landscape where the river meets the fjord. The wedge acts as a narrow promenade and the skirt that will intentionally turn into a sculptural quay wall is expected to become the main attraction within the area showing the land uplift throughout history.

The quay wall will be cast in concrete with a layering that highlights milestones in the history of Oslo. The main story to be told by the skirt is commissioned by the Oslo municipalities. Beyond the three metres of ground upheaval that has happened over the last 1000 years, they would also like a selection of historical milestones in Oslo's transformation into the city that it is today, to be highlighted in order for it to serve as a tourist attraction. The idea is to have the milestones plotted onto the three metres of soil that

Image 8.1 Graphic illustration (rendering) of Bispekilen Øst, Bjørvika in Oslo. The wedge with river in the middle encased between tall buildings and passages for pedestrians on both sides. The skirt/quay wall runs along both sides of the river above and below sea level with stairs built into it.

Illustration: SLA

make up the quay wall. Hence, elevation point -1 will represent the Viking Age year 900; elevation -0.52 will represent the Battle of Stiklestad in 1030 during which King Olaf II of Norway was killed; and so on and so forth up to, for instance, the Kalmar Union year 1624 (elevation level 0.96) when three kingdoms in Scandinavia joined forces, and later in year 1905 (elevation 2.92) when Norway became an independent country up to today's Oslo (elevation 3.46) with the urban developments of Bjørvika.

SLA has been hired to figure out how to solve the task of translating the story into material construction, and Sophie explains to me how she is approaching the task in 'the SLA-way', finding a reference point in nature. As one of the first things, in the office she scrolls down her mental map of earthen walls and then she turns to Google to expand the images of soil layers in nature. In a conversation at some point later, she shares with me her fascination for earthen walls that she senses: 'There is something seductive about seeing a cut either through a mountain, dunes or other masses of soil. There may be some knobbly transitions between the layers, and each layer of soil tells its own story of what geologically happened at this time'. The atmosphere of soil layers – the felt presence of stratifications of rocks and soil layers – envisioned by Sophie here turns into a guiding design principle at an early stage. For a start, it prompts her to advocate for a skirt surface that is rough and palpable in close-up like a real earthen wall that will also manage to 'add a tactility to the space'. Later in the process, however, the rough surface gets rejected in favour of a smoother and more 'graffiti-cleaning-friendly' surface required by the municipality. Clearly, the process is constantly bound by feasibility while simultaneously being adapted and regulated according to the guiding design principle of an earthen wall.

The nature-based design in this case does not aim to copy nature on a 1:1 scale, Sophie explains. The risk of it turning into a corny imitation of an earthen wall would be far too high, she reckons. As explained in Chapter 3, SLA's new city nature is neither untouched nature nor 'traditional' urban greening – it is a designed 'nature' that encourages its independent appearance in experience. Sophie says:

> It's our job to find out how to solve the task based on technical construction conditions. In other words, we are locked by a few constraints that involve, for example, casting in concrete and making a mould. It's expensive to make a mould, so how do we make a mould that we can repeat as many times as possible, while you still get the same experience as if you were facing a real earthen wall in nature?

The goal, thus, is to lift an experience from one setting to another – to transport and imitate only the atmosphere without transporting and imitating what originally materialised or situated it. For the skirt in Bjørvika, the ambition is to make it evoke the same kind of wonder that a 'real', untouched earthen wall would, without necessarily looking the same.

Of course, what we create is based on built premises, since it must be able to last for many years and besides it must pay off. But that's where you, as designer, need to study some different profiles [and ask yourself]: What comes out of making the shape wavy in this way? Should it instead be jagged, or should it simply have bigger waves? I need to see it from a distance and up close to evaluate it based on many parameters all the time, and the goal is to create a wall that can give a feeling of nature. It should not end up as a polished bank wall, you know.

Sophie laughs.

The groundwork in AutoCAD

In AutoCAD Sophie starts by drawing what she calls 'the principle' that is going to serve as the very foundation of the skirt. This principle will communicate about the history literally being layered into the wall, and this is where she does what she calls groundwork', calculating the shape of a structure within a grid that will quickly give her a sense of scale (see Chapter 6 on scaling). 'How big is the wall in comparison to a human figure?' she asks herself while she inserts a silhouette of a man. Where do the different elevated layers go on the wall? Where would a staircase fit into the drawing? Which part of the wall is under and above the water surface respectively? All these measurements are plotted into a template.

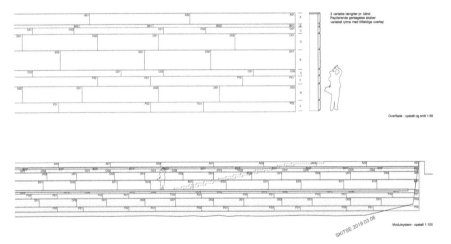

Image 8.2 A scale model (preliminary sketch) showing the module system of the skirt at a scale of 1:100.

Illustration: SLA

The scale model (Image 8.2) appears unfinished and temporary in terms of detail but is exact in its representation of the measurements for the construction. It is a provisional sketch – as a stamp says in red – and it still needs to be tried and tested. At this stage it is simply processed by an algorithm. Upon setting the first measurements, Sophie starts to test the experiential effect of the structures in PhotoShop and later she converts 2D to 3D. This 'is often the most interesting exercise; to take something as rational [constructed] as a surface or a grid can be – that is, a logically calculated pattern or structure – and then you expose it to a natural effect and then something unpredictable happens', Sophie tells me. The natural effect that the skirt drawings are exposed to comes, in this case, from daylight. Sophie needs to switch from AutoCAD to Adobe to recognise the extraordinary influence that light and shadow has on the quay skirt experience. As it turns out, the meeting between the solar/shadow algorithms and the drawn construction will come to lead the design process forward.

The building blocks are formed by the way light and constructed curves meet

The first time I talk to Sophie about this project, her work on finding the right surface for the wall has already undergone several phases. Wood cladding with vertical ebony boards has been rejected and now there is agreement that the wall should be cast in concrete. But what should the profile look like? Sophie googles for images that can give her some ideas. She types in constellations of keywords that 'bind' her. When she and her colleagues speak about bindings, they normally mean the restrictions or matters that are not up for discussion. In this case, what binds the skirt design concerns the 'façade' + 'concrete' + 'wall' + 'soil' + 'texture' + 'horizontal' + 'architecture'. Sophie finds inspiration in a wall on a random Southern European museum. She likes this wall from a distance and less, she admits, in close-up. She scrolls down for similar pictures – still on Google – and she starts to select and compare images of natural soil layers and constructed façades respectively 'that can do a bit of the same thing', as she says. The pictures serve as inspiration for her drawing of a profile shape, and at some point, after 'playing around' with different kinds of shapes for a while, she needs to test out how they allow for different 'readings' of the space within Bispekilen Øst in Bjørvika.

To 'read' space – real or virtual – is common colloquial terminology in the office. I attend to these readings as ways of tuning in on the experienced quality of spaces; when read, it seems that these designed spaces tell about their atmospheric vibe. At one point Sophie seeks to 'test out' two drawings by making two such 'readings', and I observe how Sophie starts to attend to the materiality of the design (see Chapter 6) – that is how the design 'behaves' when it meets with 'natural effects' such as sunlight.

'The pictures here [images 8.3 and 8.4] discuss whether we should go with a square shape for the louvres or a soft wave', Sophie tells me. 'Look', she carries on while she points to the first sketch (Image 8.3), 'this was in

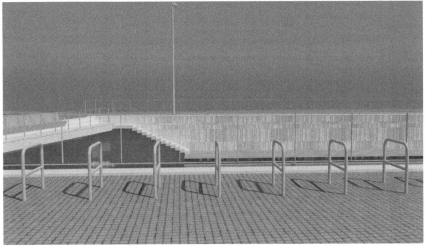

Images 8.3 and 8.4 3D graphics of the skirt surface seen at a distance, from the opposite side of the river bank. Image 8.3 (top) shows a profile with square-edged louvres in bands stacked staggered on top of each other. Image 8.4 (bottom) shows a profile with soft curves in bands stacked right on top of each other.

Illustrations: SLA

principle just stupid because the surface is basically just a smooth surface, and nothing really happens in the meeting. Somehow it only refers to the soil layer by its surface and that's it'. At this point, I am not quite sure what Sophie means by her mention of a 'meeting'. What is it that meets the surface and what does not happen? In the first image (8.3), the surface consists of louvres, and when Sophie takes note of the surface, she considers its reference to layers of soil that apparently, in this image, *solely* happen by virtue of its surface. Hence, there seem to be other parameters of reference that she is missing. It may well be that the louvres communicate visually about soil layers, but apparently this communication does not satisfy Sophie. It is too mono-sensory, it seems.

'It's merely smooth and therefore it's not particularly exciting to touch', Sophie then complains with an indication that a tactile appeal is still missing. The meeting apparently involves more than what meets her eye. It is expected to hold a haptic, palpable and touchable feel comprising her imagined hand touching its surface. When Sophie examines the wall's 'behaviour' in the images, she continuously needs to change her perspective: 'I have to get close and far away', she says. Attending to the wall surface up front, she imagines what it must be like to run a hand over the uneven but smooth surface of the louvres as if she were reading Braille, and from a distance she observes the overall impression given by light. This is where the meeting happens, but it is not only at close quarters that the wall can be touched. It also touches from a distance. Below I will illustrate how the visual impression given by light and shade dancing on the wall observed from a distance has a tactile power which is not only tested by Sophie's imagined hand 'reading' its surface, but a tactile force that draws her into the drawing.

In Image 8.3 Sophie sees 'that the square profile of the louvre results in a rigid wall that casts a comb-shaped shadow which makes it almost holy and solemn'. The wavy surface (Image 8.4), by way of contrast, behaves differently; it makes the wall soft and exciting because it casts an 'infinitely different' light. Sophie's emotional reading of the wall that becomes either holy and solemn or soft and exciting tells of her emotional commitment to the design given by the meeting between sunlight and the shape of the logically calculated construction that becomes decisive for the further development of the skirt design. The meeting that takes place virtually between the 'forces of nature' in the form of sunlight and 'built' structures in the form of profiles is no longer just something Sophie is drawing or that she works out while she imaginatively touches it. It also draws her in: 'You feel like touching the wall, just like when you stand in front of layers of soil that are millions of years old and you just *need* to touch it', Sophie says. She emphasises this need with a rising pitch in her voice. Although the louvres in the first version of the skirt make an uneven surface that she likes, they still create a rather 'square rhythm' which is monotonous when viewed from a distance, Sophie says. The contours of the wavy profile (Image 8.4), however, make the wall lighter and softer, and thus 'it gets friendlier', less pompous, and has a 'completely different rhythm', Sophie clarifies. At this point in our conversation, I notice how Sophie might literally be 'seeing rhythm', and I therefore ask her what she means by rhythm. Here is her response:

It is the light that is the rhythm and the shape. Where I had made it too much of a stepped gable, large shadows where projected. But where I drew soft curves, it became lighter. Or... vice versa, perhaps: where there was light, it became soft. Whatever, I got some surfaces that looked very flat and bright. The light is constantly present. When I speak rhythm, it is the light that makes me do so.

The interplay of light and shadow turn out to play a crucial role in the design process, but what is decisive is how it *meets* the construction. The square profile shape (Image 8.3) clearly does not generate the same lively shadow play as the shape given by Image 8.4 with the wavy profiles. 'Wavy profiles make the wall appear much more exciting and unpredictable', says Sophie. Of course, there are always practical circumstances as well to consider when decisions are to be made about construction work. In this case the square profile of the louvres is, with its 'perfect shape', vulnerable to wear and tear because fractures are easily seen as damage, whereas the soft and irregular shapes react to wear differently – tending to turn damage into the *wabi sabi* beauty of patina (Chapter 3). Yet it is only when Sophie exposes the drawings to light conditions via PhotoShop that she can make sense of their materiality (see Chapter 6) – what they do to the space in terms of its felt quality. Light was already conceived as a building material in line with steel, concrete, glass, etc., within architecture (Bille and Sørensen, 2007), and the current study is therefore clearly restricted to ethnographic findings that show how light and lighting alone seldom determine design choices, but rather it is *light in interplay* and correspondence with materials, angled, shaped and drawn by a designer, that takes the design further.

Deep shadow drop forces a hierarchy onto the basic narrative of Oslo's history

Image 8.3 show that, at some point, Sophie and her team were working with displacements of the horizontal bands that refer to soil layers. Displacements in depth appear like the bands being pushed in and out between each other. In the cross sections illustrated in Image 8.5, we see how a first layer is pushed out, the second layer is pushed in, the third layer is pushed out, and so on and so forth. Originally, Sophie had made a square profile with bands (the layers) stacked right on top of each other and no displacement in depth (Image 8.6), but the design process has taken many twists and turns along the way and alternative solutions are constantly introduced, tested and discussed back and forth. She does not quite remember the reason for introducing displacements – it was probably a developer's clever idea that made her draw them, and then later as it was tested, it would turn out to be a less clever idea, Sophie suggests. Hence, in the basic AutoCad layout the lines for a structure with displacements had probably just been drawn quickly in order to satisfy a client's

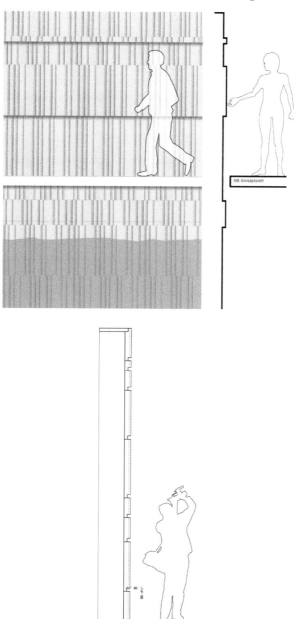

Images 8.5 and 8.6 Cross-section of the skirt versions. Module structure with displaced bands/layers in depth (Image 8.5 on the top). Square profile with bands stacked right on top of each other with no displacements (Image 8.6 below). The first and final solution is below (Image 8.6).

Illustrations: SLA

immediate wishes until, at a later point, more in-depth investigations in PhotoShop would question their suitability.

One day, however, as Sophie finds time to try out the drawings in 3D (Images 8.7 and 8.8), it becomes clear to her that the displacements are no good for the construction. In fact, it turns out that they disrupt both the basic narrative of the wall and SLA's nature-based design principles. As for the disruption of the main narrative, the displacement of layers running in and

Images 8.7 and 8.8 3D graphics of the skirt surface, both with soft wavy profiles and seen from the side. Image 8.7 (top) shows a profile with two dominant no dominant heavy shadows from some bands stacked right on top of each other. Image 8.8 (bottom) shows a profile with no dominant heavy shadows from some bands stacked right on top of each other.

Illustrations: SLA

out causes two edges representing two out of several milestones in Oslo's history to cast significant shadows, and all the while other milestones (and edges) are not able to throw a shadow (Image 8.7). Sophie says: 'When the sun gets very bright, two elements stand out very noticeably'. The first and fourth layer thus emerge with shadows that seem to make other milestones less visually relevant. Hence, the two big shadows highlight two milestones at the expense of others, and as a spectator one will inevitably ask oneself why two years are more important than the rest. The truth is that no milestones should be highlighted as more significant than the others, but the shadows that are dropped from the structure with displaced layers moving in and out in depth obviously introduce an undesirable hierarchy into the narrative of Oslo's history.

Sophie examines the wall with displacements in depth and tests how it appears with square and curved profiles in the wall surface, respectively. It does not make much difference either way. And not only does the displacement interfere with the dissemination of Oslo's historic milestones, also it disrupts the nature-based design, through which SLA seeks to promote systems of nature as opposed to structures of construction – non-hierarchical order in contrast to neatly sub-divided hierarchies of construction. Sophie teaches me that the displacements, with their heavy shadows, unintentionally promote 'everything we try to hide or disturb – that is, the fact that the skirt is made of built elements that are hung up on a wall, suddenly becomes enormously clear and prominent due to light and shadow'. This goes against intentions that seek to get rid of the built construction in experience – not in material reality but as a feeling. Because of the uneven occurrence of the two huge long shadows, the wall is no longer experienced as a whole but is instead experienced in broken built elements arranged in a hierarchical structure 'just like the block of flats that surrounds the wedge', Sophie says, referring to the logical construction of the surrounding built environment (see Image 8.1, far left and far right). When the displacements are removed (Image 8.8), in turn, the wall loses its hierarchical structure of elements and is again experienced as a natural whole, and all the while Oslo's milestones are given the same degree of attention. In fact, the milestones go from being communicated in the concrete structure to being told by the shadows, and it is this subtle finesse of creating the intriguing shadow play that Sophie handles in that part of her work when she describes experiencing it as a whipped cream effect, as we see below.

Shadows communicating history by dancing like whipped cream

For quite some time, Sophie had a hunch that the curved profiles could offer something that was not quite clear to her yet. At some point, two and a half years into the process, she is finally allotted the time to 'nerd through' the 'investigative drawings', as she calls them. In her nerdy investigative work, she is particularly interested in making what she calls a 'random system' that, when the concrete elements are stacked directly on top of each other and not shifted in depth but only pushed from side to side according to a random system calculation, the shadows that appear right where the layers meet start to behave like whipped cream. 'You know', she explains, 'when you whip the cream and at some point, it starts to make these soft, fun shapes that are completely random'.

It was an intuitive guess rather than logical consideration that finally helped Sophie draw the random whipped cream effect. 'When I tested it in 3D, something just completely unpredictable and funny happened in the overlaps, and then it dawned on us that it is precisely the overlap and the meeting between the bands stacked on top of each other which is interesting. That is, the overlap that we until this point had tried to hide by pushing the layers in and out between each other. Where the displacements, described above, pushed the bands in and out in depth, the pushing of elements back and forth horizontally instead – a bit to the left and a tiny bit to the right – made all the difference.

The way the shadows cast from the edges of each band started to behave like whipped cream in curly transitions is, says Sophie, 'a bit like what real soil layers do when you touch them. In layers of soil there may be a rock or a stone sticking out here and there and the tactile impression makes the experience 'flawed'. 'That's actually what it's all about', says Sophie. Of course the whipped cream shadows cannot be touched like small rocks and stones can, but the edges and the curvy profile of the skirt is to be touched just like stones can be touched. What Sophie seems to indicate is that she is well pleased with the way she and her colleagues managed to 'translate' the haptic and visual sensation of an earthen wall into their skirt. Images 8.8 and 8.9 illustrate

Image 8.9 Whipped cream showing random curves and shadows similar to the desired materiality of the skirt in Image 8.8.

the different forms and shadows of different materials that Sophie neverthe-less connects due to their experienced materiality. The unequal, flawed and deviating aesthetics of an earthen wall she captures as an atmospheric mate-riality given by light and particularly shadow dancing on the skirt cast in concrete which, in experience, begins to behave like whipped cream (Images 8.10, 8.11, 8.12 and 8.13). The irregular shape of the shadows is essential to the atmosphere that begins to emerge around her skirt experience. And the shad-ows occur partly due to her removal of the displacements in depth (described above) and partly due to her addition of horizontal displacements, pushing the concrete blocks in unequal measure and with what she calls a 'chopped sine wave 'from side to side in between each other in layers.

When the shapes are mounted in sections of horizontal bands, as if they were curtains, their chopped sine waves make them always meet each other differently. So instead of each curtain folding in a uniform way according

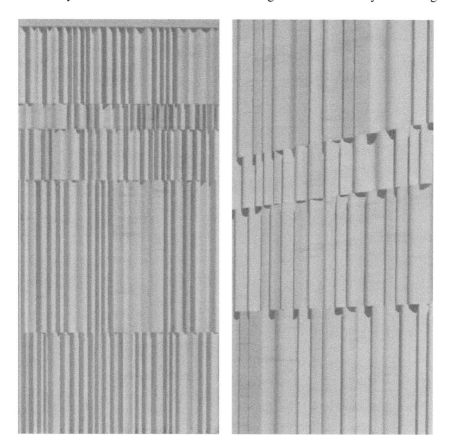

Images 8.10 and 8.11 Visualisations of the skirt profile: from the front (Image 8.10, on the left), from the side (Image 8.11, on the right).

Illustrations and photo: SLA

Images 8.12 and 8.13 From the side in close-up (Image 8.12, on the left). Next to the profile images is a reference photo (Image 8.13, on the right) of calcareous soil layers showing the imitative design's reference to soil behaving like whipped cream.

Illustrations and photo: SLA

to the curtain of the band just above or below, Sophie makes sure that each curtain will never reiterate the folding of a neighbouring curtain. It is this 'random system' of how the curtains fold out of sync with each other that produces the 'whimsical shadows' that Sophie compares with how whipped cream behaves.

Building on the previous Chapter 7 on the mimetic design faculty, we see here that atmospherically the bands with wavy profiles start to *engage with* each other when they are stacked and when sunlight *meets* their surfaces, which resembles how 'undesigned' soil layers behave atmospherically when their uneven surfaces made of clay, stones and rocks are met by light rays 'in nature'. And so too does the agitation of air bubbles added by whipping them into a bowl of heavy cream *engage with* the fat globules that together transform when they begin to coalesce in chains and clusters and spread around the air bubbles. What Sophie seems to suggest is that while the three physical events maintain material characteristics that objectively are very different from each other, they still strike up the same kind of astonishing atmosphere do to their materiality that fascinates due to its wonder. From the skirt with levelled and sideways shifting layers 'you get something unique', says Sophie. 'The overlap is different in each place and therefore unique and so you may want to see the whole wall or take a closer look'. So while Sophie engages in potential user experiences, she is also driven by her own aesthetic fascination with the potential of uncertainty, which is clearly played out in the whipped cream effect.

Design as a process of correspondence and anticipation

A theoretical counterpart to the nature of Sophie's investigations, which witness a hodgepodge of connections that permeates materials and materiality of the design, is to be found in Ingold's concept of correspondence, that describes a relation to the world (2013: 7). As mentioned in the earlier chapters, correspondence helps to see not only how materials bend and twist in correspondence with each other, but also how designers such as Sophie react to what the material interplay can offer to the design process in terms of their materiality (Chapter 6). Ingold advocates the *morphogenic* (form-generating) approach to craftmanship and design production in favour of the *hylomorphic* approach, from *hyle* (matter) and *morphe* (form). Hence, instead of following blind the customary belief that the creation of artefacts – such as making a skirt in Bjørvika – starts with an idea that achieves its form by having a material supply followed by a moulding activity (Ingold, 2013: 20), the form-generating approach instead helps to see Sophie and her colleagues in a much more humble role, engaging and co-designing not only with each other but clearly also *with* materials, their materiality and light and shadow, through drawing and image making. On Ingold's advice: 'even if the maker has a form in mind, it is not this form that creates the work. It is the engagement with materials. And it is therefore to this engagement that we must attend if we are to understand how things are made'. (2013: 22). It may be this engagement that Sophie unwittingly refers to when she says that the design development is all in the process:

'I didn't know in advance that this was how it should be solved'. Sophie sighs in relief as the whipped cream effect of the skirt is revealed to her. 'It's not that the new solution is any more difficult or complicated to make, but I just couldn't work it out from the start. It just takes time. It's a process. And each time it's about finding out what is possible and then testing out how it works'. And as she was to show, the final 'whipped cream solution' would prove to be much simpler than the complex staggered, 'holy' and hierarchical alternative. For instance, the rejected displacements in depth were examples of a design that was too heavily laboured – like in the case of the monkey cave described in the Chapter 7 – that caused too complex a level of detail, requiring too many drawings to be made without any gain. What we learn from Sophie seems to confirm that the idea of hylomorphism is rather misleading. She could not have figured it out in advance, she said – hence, she was not moulding matter into a prefigured shape.

Occasionally, throughout the process and while I shadow her at work, I sense that Sophie might even be taken by surprise now and then. I therefore ask her: 'Are you occasionally taken by surprise, do you think?' And Sophie answers:

> Yes. I guess so. I mean, it would be really boring if I could tell from the start how things would be. If I could just decide everything independently and allow my own personal taste to dictate how Bjørvika should look, then, in a way, there would be no need for me to involve myself as a designer. It would be much easier to select standard items from a catalogue and patch up the area with prefabricated items that would make any corner of Bjørvika look the same.

If we take Sophie at her word, we are not to understand the designer as a person who can tell what will come out of the design process beforehand. Nevertheless, this might be what many developers seem to expect from designers. It seems almost a prerequisite to the creative process, here, that the designer is never ahead of the design's development but *with* it. In Richard Sennett's work on craftmanship we find a discussion about what anticipation means to processes of making. Anticipation rightly implies, Sennett writes, always being 'one step ahead' while working with materials (2008: 175), but it does not for that reason involve preconception. I read this as a suggestion that there is a temporal attitude that is neither in the present nor in the future; not with two solid feet fixed in a present moment, a little too laid back yet open to what shows up, but rather in a dynamic and alert position with one foot in the present, and one foot lifted ready to take whatever shows up the next step forward. Ingold has addressed this kind of anticipation, ahead of the present moment, as 'foresight', because it describes a way to 'see forward' not through 'cogitation that literally comes *before sight*' as when 'procedures [are] precisely laid down in advance' but through a bodily way of engaging with the materials as they transform the design process (Ingold, 2013: 69).

Revisiting Sophie's work with this conceptual discussion in mind, there is clearly no sign of her projecting ideas onto future situations. At one point she even says:

> You cannot imagine everything. That's what it means to be in a creative process; it means to sit and try out, test and examine.

And her examinations involve her empathic engagement in how meetings unfold between structures of lines, light and shadow, hierarchies, narratives, etc.

On the other hand, the drawings do not surprise her out of the blue. It is due to her exchange *with* the drawings that she finds herself *within* the creative process crucial for the design development. The atmosphere approach that I advocate here is therefore less interested in power and agency, and accordingly less occupied with dilemmas 'between the self that is "in control" and one that is "led"' by others (Yarrow, 2019: 112). Rather than deconstructing, separating, and hierarchising, the atmosphere approach seeks ways to converge, merge and search for connections that dissolve 'selves' and 'others'. Ingold suggests thinking of drawing 'not as the projection of ready-made image but as the inscriptive trace of a movement or gesture comparable to weaving' (2013: 71). In parallel with Yarrow's identification of the architectural response to a place as a matter of *unfolding* buildings from places rather than fitting buildings into them (2009: 79), I have a strong sense here of the Bjørvika team unfolding the construction of a skirt from drawings that involve a reciprocal relationship between the designer and 'natural' and 'constructed' event as they *meet* virtually.

Drawing atmospheres in resonance

Ingold's conceptual understanding of correspondence in design processes offers a fruitful approach to the study of atmosphere in these processes, because it enables a sensitive approach to how bodies of a human and non-human kind draw in and move each other in cohesion. It allows us to see how lines that are drawn can bend the light, and how light produces shadows that dance in correspondence with shapes of profiles cast in concrete. In comparison to Science and Technology Studies and Actor Network Theory, it is a perspective that offers to consider the feelings that are involved when bodies meet and how they transform each other in the way we have seen in Sophie's drawings of a skirt. With reference to a string-making exercise performed with a team of students, Ingold calls attention to the feelings involved in this task:

> The strands of our string, as they twisted around one another, had a feel for each other in their correspondence no less than that of the eye-beams or heartstrings. They are but species of the same phenomenon. The language of feelings is appropriate, and as literal, for the twisted eye-beams of lovers as it is for the entwined strands of strings.
>
> (Ingold, 2013: 121)

It is, however, not just materials such as strings but also, as in the case of the skirt, the materiality of lines, solar/shadow algorithms, etc. that twist and link emotionally. The designer – in this case Sophie – engages in the drawing atmosphere as well and she constantly tests and evaluates what she takes to be, for instance, a soft, 'holy' or whimsical materiality of the skirt.

Ingold quotes John Berger in saying that in the state of 'drawing-thinking you become what you draw: not in shape but in affect' (Berger, 2005: 126). Ingold elaborates that, when 'draughtsmen' draw, they will 'know it from the inside, and in [their] gesture [they] relive its movements' (2013: 129). This way of literally knowing drawing by the way that it issues from the body has much in common with how Hermann Schmitz thinks about atmosphere being bridged from 'incorporeal things' to human bodies through suggestions of movements (Schmitz, 2009: 33). The reason why it may be necessary to adjust and supplement Ingold's terminology and conception of correspondence is because it suffers from a material limitation.

Sophie sees, imagines and feels so many things that one almost forgets that it all happens in front of a computer screen. In some projects, occasionally mock-ups are built, but not in this case. There is no physical skirt yet – the skirt is on its way and its genesis process is bound to digital modelling. But computer-generated drawing challenges some of the fundamental ideas that support Ingold's thoughts on correspondence. Initially Ingold addresses material making exclusively and sees the tactile handling (that is, the touch) of materials – including sketches in terms of drawings made by hand – as crucial to the designer's feel for the design – and the potential of the designer to not only touch but be touched mood-wise by the design. Ingold writes:

> At the point where the sketch gives way to technical drawing, all movement is stilled. The lines of the technical drawing may encode instructions on how to move, but convey no movement in themselves. And for the same reason, such movements are devoid of feeling. They establish a relation with the world that is optical rather than haptic.
>
> (Ingold, 2013: 126)

This narrative is not consistent with the reality I meet in SLA during my fieldwork. I therefore argue, instead, that lines of technical drawings *do move* and they are not devoid of but charged with feelings in situations when they meet other constituents of the design. Nothing stands alone but moves together. Everything is inevitably in atmospheric encounters. In order for Sophie to feel the drawings, she needs to produce 3D images. She is not able to feel the drawing's atmosphere of the skirt either in hand-drawn sketches or in technical drawings alone. Although technical drawings play a crucial role (they are part of the 'groundwork', as Sophie says), it was still the graphic software that enabled Sophie to feel the drawings due to its 'gathering' contribution of bringing together the many constituents and the materiality of the design.

Because I stretch Ingold's conceptualisation as far as is the case here, it might be advantageous to think in terms of resonance instead of correspondence. Resonance is a much broader term that can be defined as 'a kind of relationship to the world [...] in which subject and world are mutually affected and transformed [...] not as an echo but as a responsive relationship" (Rosa, 2019: 174). We can think of the designer as the subject that meets the drawing and instead of a reciprocity in 'affect' or 'effect', I propose considering drawing as a crucial design activity that draws in a double sense: as a process of image-making and as a way to relate where the drawing draws in the designer while the designer draws up the details, tests, evaluates and refines until a satisfying result is reached. I understand resonance via Böhme's conceptualisation of atmosphere as a way of oscillating (*mitschwingen*) with one's surroundings. He writes: 'In perception, one's own existence becomes noticeable through the fact that one is hit, that one is exposed to the world and resonates with what is perceived' (Böhme, 2021: 83, my translation). Related to Sophie's drawings as described above, we can see how she resonates with the drawing and how it develops as she allows it to 'touch' her. Different conceptual usages allow us to look at the situation differently. Resonance is not defined as a specific 'emotional state, but a mode of relation that is neutral with respect to emotional content' (Rosa, 2019: 174). It therefore allows for any feeling – any atmosphere – to appear in any situation. While resonance addresses 'the relationship [that] precedes that which is related' (Rosa, 2019: 383), atmospheres, moods and feelings describe the resonant qualities in experience.

In order to see how Sophie resonates empathically with her drawings, it may be helpful to draw on Schmitz's notion of suggestions of movements defined as bridging movements where atmosphere can travel from 'incorporeal things' to human bodies (Schmitz, 2009: 33). I tend to think of suggestions of movement as embodied empathy that enables the encounter of atmosphere in our surroundings. It is argued by Schmitz that suggestions of movements incorporate what exists and remains without bodies. Hence, he offers an approach that attends to the way emotional incorporation succeeds not only among human bodies – we can think of the mirror-neuron mechanism as an example – but also among bodiless things and events and corporeal human beings. Atmosphere is encountered via such corporeal movements (Schmitz, 2014: 5). In comparisons already described between an image of an earthen wall, profiles to be cast in concrete and whipped cream, I see how Sophie fastens on a specific kind of atmosphere. The bodiless things seem to suggest an aesthetic where elements of uncertainty play important roles. When they resonate with Sophie, their unpredictability is embraced as it marks her mood. This appealing atmosphere that plays with elements of uncertainty is bridged again and again, also when Sophie tries to describe to me the atmosphere of the skirt during an interview. She does not express herself in words but asks to borrow my pen and notebook and then starts to draw by hand two lines that intermingle – a bit like Ingold's strings twisting around each other (Image 8.14).

As she draws the two lines, her voice starts to assist the curvy movements of her hand, and I realise that with her voice she is mimicking what seems to be an irregular rhythm – the arrhythmia of the 'random system' she was searching for throughout her investigative drawings. This was how I came to learn about Sophie's way of drawing the atmospheric skirt not only through

Image 8.14 Architect Sophie's drawing in my notebook showing two lines that intermingle. The drawing is made while she explains to me how she experiences the whipped cream effect of 'whimsical shadows' from bands that
meet each other randomly.

Photo: Anette Stenslund

my visual observation and intellectual conversation with her and others, but
also by developing an ear for the beat of her drawn lines that I would come
to feel myself to understand it.

Conclusion

In this chapter, thorough observations of how atmosphere from a designer's
perspective is drawn into future space illustrate how drawing itself pulls
constituent bodies into the design process. Digital drawing of urban design
is demonstrably shrouded in atmosphere. It is not projected from a designer's predetermined mind onto the design, and it is not sent out by things'
ecstasies onto a receptive designer. By pursuing the process of creating a
skirt for Bjørvika, Oslo, the chapter goes into detail about the atmosphere
that envelops the designer's way of resonating with the materiality of many
parts of the design – including the design narrative, geometric measurements, shapes of profiles, light and shadows. The drawing process is characterised by a form of correspondence, but the correspondence transcends
a physical and material exchange. It is characterised by a felt and sensed
relationship best grasped as resonance loaded with moods as atmosphere.
Drawing atmosphere happens in atmospheres that pull its constituents into
cohesion in virtual 'meetings'.

Focusing on the processes of drawing, it becomes clear not only that
the design takes shape through countless negotiations and practical considerations but also that drawings for construction due to their materialities act as crucial co-creators of the design. Accordingly, the chapter
has shown how the urban designer cannot rationally imagine the right
atmospheric solution before it develops and reveals itself through the
empathetic way of drawing, testing, sensing and feeling. As demonstrated
through Sophie's observance of principles, systems and patterns that
emerge in, with and from her drawings through her investigative work,
we see the added value of a designer's attitude of resistance against overthinking and oversteering the design phase in detail. The benefit of a

designer's attitude that remains sufficiently open to let them show what they 'couldn't have thought of' intellectually beforehand is self-evident. It has been argued that anticipation is an essential part of craftmanship, and this chapter has drawn the reader's attention to ways of anticipating atmosphere in drawing.

References

Berger J (2005) *Berger on Drawing*. Cork: Occasional Press.

Bille M and Sørensen TF (2007) An anthropology of luminosity: The agency of light. *Journal of Material Culture* 12(3): 263–284.

Böhme G (2001) *Aisthetik: Vorlesungen über Ästhetik als allgemeine Wahrnehmungslehre*. Munich: Wilhelm Fink.

Chumley L (2016) *Creativity Class: Art School and Culture Work in Postsocialist China*. Princeton: Princeton University Press.

Cuff D (1991) *Architecture: The Story of Practice*. Cambridge: MIT Press.

Ingold T (2013) *Making: Anthropology, Archaeology, Art and Architecture*. London: Routledge.

Rosa, H (2019) *Resonance: A Sociology of Our Relationship to the World*. Cambridge: Polity Press.

Saint A (1983) *The Image of the Architect*. Wallop: Yale University Press.

Samuel F (2018) *Why Architects Matter: Evidencing and Communicating the Value of Architects*. London: Routledge.

Schmitz H (2009) *Der Leib, der Raum und die Gefühle*. Bielefeld: Sirius Edition.

Schmitz H (2014). *Atmosphären*. Freiburg: Karl Alber Verlag.

Sennett R (2008) *The Craftsman*. London: Yale University Press.

Yaneva A (2009) *The Making of a Building: A Pragmatist Approach to Architecture*. Bern: Peter Lang.

Yarrow T (2019) *Architects: Portraits of a Practice*. Ithaca: Cornell University Press.

9 CONCLUSION

Atmospheres are inherent to human life; they are always already with us – and inevitably, as I have shown, they are with the urban designers in their working day. Their presence is unquestionable, yet atmospheres tend to flee from any attempt to capture or get hold of them. Occasionally they are verbalised, but words may not fully capture them. They are in a constant flux and remain inexhaustible to any designer's full understanding. Thus, they cannot be designed according to a predetermined plan. Atmospheres do not give in to the will of designers. As shown, designers' attempts to cut through atmospheres by detailed design approaches are likely to end up with little success. But that does not mean that designers can refrain from dealing with atmospheres, either. Atmospheres in redeveloped areas may be held against the designers should post-occupancy evaluations show low user satisfaction. Hence, rather than trying to control or escape them, this book suggests that designers profit from their potential – which requires that we are able to spot them. This book makes the case for a deeper understanding of how atmospheres are not simply end products of design businesses but that they envelop processes of making and that they are vital to the genesis of design.

Based on ethnographic fieldwork among urban designers in SLA that includes observational and re-enactment studies, office-located and on-site walk-along interviews in urban areas for redevelopment, archive studies of text, drawing, image-making and literature reviews, the most overarching observation is that urban design feeds on atmospheres from *within* the very process of making. And second, more controversially, it shows that designers do not only design for others but for themselves to arrive at a common course within their teams of collaborators and to lift their standards of innovation. Atmospheres are the glue, you could say, that holds together the cooperating bodies and ensures progress in the processes of development and execution of projects.

Urban designers engage in atmospheres in order to design their products and services. For example, they depend on resonance with their drawings and various kinds of CGI in order to feel, sense and bodily experience the quality of the design. At SLA this was explicit, but there is nothing to suggest that this may not also be the case elsewhere in more

DOI: 10.4324/9781003279846-9

implicit forms. As it turns out, the types of appealing designed atmospheres often emerge from situations where uncertainty is embraced rather than eliminated and turned into certainty with answers and predictable results. Atmospheres of uncertainty create momentum in both creative, aesthetic and social situations of negotiation, and they are characterised by an attitude of the designer – a humble attitude, if you like, that does not force its design into a form that is not already proposed by the material or the materiality, in German understood as *Stofflichkeit*, of the design itself.

In design processes that embrace moments of uncertainty the designers approach their work in a symmetric power relation with materials and materiality. Atmospheres *in* urban design are thus not primarily situated in interpersonal affairs, but envelop relationships between humans and non-human bodies, including biotic and abiotic matter, that interweave, exchange and collaborate in ways that mark the bodies mood-wise. That is, designers resonate with each other and collaborators as much as with digital images, software, screens, mouses, iPhones, light and shade, colours, sizes, shapes and proportion, animals, insects, plants, fungi, buildings, wind and weather, mock-ups, pen and paper – a conglomerate of multispecies appearances that resound in the felt bodies of designers as atmosphere. Approaching urban design through atmospheres thus enables the recognition that co-designing entails exchanges of feelings that *move* bodies in two senses of the word: as motion and mood.

This book has unveiled some of the everyday situations of urban designers that inevitably 'tune' their 'engine room' atmospherically. The focus has been on the 'studio atmospheres' that appear vital to the designers' performances even if they might pass a client by. Atmospheres of uncertainty, for instance, are met by rough, inquisitorial collaging carried out for epistemic reasons that are to inform and develop a project's conceptual foundation. However, the purposive work with atmospheres that takes place, for instance, in collages, preparation of 'atmosphere sheets' and various types of drawings is marked by an outside world that does not always consider atmosphere as a valid basis for making decisions and acting. Therefore, these parts of the design work are often kept 'close to the body'; they are carried out in-house, inaccessible and remote to the public.

In SLA's office I was given a desk placed just next to a display board with a pinned printout of a quote that said: 'We know we've been innovative when we don't win competitions' (Image 9.1). This was how SLA was quoted in a special issue of Vision Magazine, China's most influential and trendsetting lifestyle magazine. Since it accompanied me during the fieldwork, it feels that this quote never really stopped teasing my mind. To be innovative means to come up with a completely new idea or vision – or to implement an idea that gives rise to a new process, product or service for commercialisation. Given that anything new is not yet known, it is reasonable to expect that innovation entails uncertainty. What SLA seems to suggest with the quote, however, is that the outside world reacts dismissively to innovative

Image 9.1 Bulletin board next to my desktop space in SLA.

Photo: Anette Stenslund

proposals that engage in uncertainty. In an interview, Stig L. Andersson tells me about experiences of clients' inability to imagine or comprehend potential scenarios proposed by SLA – and if clients do not understand or see the point, they will lose tenders. There is thus a mismatch between ambitions and vision for urban design on the one hand and fulfilling clients' requirements. The question is whether knowledge of atmosphere can help here.

When designers form their impression of places and images, they draw on their skilled vision, which involves a multisensory imagination of felt sensations that no outsider is expected to master. This might be the reason for the outside world's lack of understanding of designers' work with atmosphere. For the outside world – not having a share in the SLA DNA, not being part of the studio's work culture and thus not having incorporated the same practical sense and bodily knowledge of atmospheres as have the experienced practitioners – addressing atmospheres can seem 'fluffy' and selective, even too artistic and poetic. But what happens internally in these situations is that designers engage in uncertainties for as long as the situations, budgets and timeframes allow: they do not seek answers but ask questions, test and reorganise, zoom in and out, try out in 2D and 3D until they intuitively *feel* that the design is on track. Without these feelings that rise from oscillating subjects and objects and that ooze through meeting rooms, computer screens and various types of CGI as what I identify as atmospheres, there would be no design service that could not be taken over by a robot. And it even turns out that skilled vision is not necessarily shared by all practitioners within the same industry that extends across national borders. They are tied to the corporate culture, and work done with renderings bears witness to how intuitive flair is not easily translated across collaborators from related companies.

The book has posed questions about the *feel* of urban design, both in the sense of the urban designer's imagined feel of future redeveloped spaces and in the sense of how the design development itself resounds in the feeling bodies of the designers. It shows that designers do not only design for others: they not only seek to produce atmospheres in urban space but also design for the designers themselves in order to feel its quality and to arrive at a common course within their teams. Hence, they engage in, with and through atmosphere in order to lift their standards of innovation. Accordingly, the book has discussed how urban designers are emotionally committed to one another and the design process as such, and their incursion into atmospheres is an indispensable part of the design process. The conceptual framework developed and discussed in the book serves to consider and address such felt matters anew. However, it says little about what city dwellers, clients or stakeholders feel or think about atmospheres in urban design; it talks about what urban designers believe they know about how others might feel and, to an even greater degree, it looks at how the designers' themselves feel about their work. How designers engage in atmospheres to develop their design is likely to rub off on how they perform, take decisions and enhance their services and, ultimately, how they manage to capture their market.

Imagine what it would be like if juries validated architectural firms in proportion to their ways of engaging in the design processes while of course fulfilling their obligations to their clients and to society. What if there were front-page news to be gained from clarifying procedural skills, empathic responsibility and responsiveness? This would be about what is felt, rather than measured. What if clients were content to know about designers' thoroughly skilled approaches to work – their methodologies, attitudes and ability to meet uncertainties and handle risk in productive ways – instead of expecting the end-results to be exactly predetermined before the task is even begun, knowing well from experience that there is always a gap between projection and result? Economic calculation, scientific measurements, isolated and standardised test trials are methods well applied in order to understand everything from particle pollution, material consumption, to physiological reactions and crowd behaviour. These are component parts of late modern reality that either call for or allow quantitative methods. They are needed, no doubt. Some component parts, however, are of a more complex and culturally situated nature and call for other methods. These parts of reality often concern how we feel about something – here and now and somewhere in a potential future. But to me it appears that stakeholders who yearn for certainty and strive to turn 'soft values' into 'hard facts' risk overlooking how people might feel. If it is true that scientific 'facts' are in demand to ensure that urban designers perform according to 'objective' standards of measurement and to ensure that they do not get caught up in emotions, this proves how poorly the activities of urban designers, including architects, landscape architects and planners, are understood – not to mention the users of both outdoor and indoor environments.

This book has argued that atmospheres are neither to be escaped nor fully controlled but they can be approached with open attitudes and serve as a quality stamp of urban design. While stakeholders may be convinced that it is only numbers that matter, it is the atmospheres that sell. The fiercely inquisitive atmospheres of collages, rendering atmospheres, scaled atmospheres, imagined and sketched astonishing atmospheres that appeal, attract and allure are the enterprise of urban design. And who knows, perhaps even within the processes of producing and addressing numbers, atmospheres may have a central role to play.

Index